# AS/A-LEVEL YEAR 1

## STUDENT GUIDE

## WJEC/Eduqas

# Biology

Biodiversity

Physiology of body systems

Andy Clarke

**HODDER**
EDUCATION
AN HACHETTE UK COMPANY

This guide has been written specifically to support students preparing for the WJEC/Eduqas AS and A-level Biology examinations. The content has been neither approved nor endorsed by WJEC/Eduqas and remains the sole responsibility of the author.

Orders: please contact Hachette UK Distribution, Hely Hutchinson Centre, Milton Road, Didcot, Oxfordshire, OX11 7HH. Telephone: +44 (0)1235 827827. Email: education@hachette.co.uk. Lines are open from 9 a.m. to 5 p.m., Monday to Friday. You can also order through our website: www.hoddereducation.co.uk.

© Andy Clarke 2015

ISBN 978-1-4718-4405-8

First published in 2015 by
Hodder Education,
An Hachette UK Company
Carmelite House
50 Victoria Embankment
London EC4Y 0DZ
www.hoddereducation.co.uk

First printed 2015

Impression number 5

Year 2022

Cover photo: skampixelle/Fotolia; p. 18 Dr Gladden Willis, Visuals Unlimited/SPL; p. 28 Biodisc, Visuals Unlimited/SPL

Typeset by Integra Software Services Pvt. Ltd, Pondicherry, India

Printed and bound by CPI Group (UK) Ltd, Croydon, CR0 4YY

Hachette UK's policy is to use papers that are natural, renewable and recyclable products and made from wood grown in well-managed forests and other controlled sources. The logging and manufacturing processes are expected to conform to the environmental regulations of the country of origin.

MIX
Paper | Supporting
responsible forestry
FSC™ C104740
FSC
www.fsc.org

# Contents

## Content Guidance

| Topics | WJEC AS | Eduqas | |
| --- | --- | --- | --- |
| | | AS | A-level |
| The evolutionary relationships between organisms  6<br>    Biodiversity . . . . . . . . . . . . . . . . . . . . . . . . . . . . . 6<br>    Classification . . . . . . . . . . . . . . . . . . . . . . 10 | Unit 2 | Component 2 | Component 2 (Continuity of life) |
| Adaptations for gas exchange . . . . . . . . . . . . . . . 15<br>    Gas exchange surfaces . . . . . . . . . . . . . . . . . 15<br>    Gas exchange in mammals . . . . . . . . . . . . . 16<br>    Gas exchange in other terrestrial animals . . . . . 19<br>    Gas exchange in fish . . . . . . . . . . . . . . . . . . 20<br>    Gas exchange in plants . . . . . . . . . . . . . . . . 23 | Unit 2 | Component 2 | Component 3 (Requirements for life) |
| Adaptations for transport . . . . . . . . . . . . . . . . . 27<br>    Transport in plants . . . . . . . . . . . . . . . . . . . 27<br>    Transport in animals . . . . . . . . . . . . . . . . . . 39 | Unit 2 | Component 2 | Component 3 (Requirements for life) |
| Adaptations for nutrition . . . . . . . . . . . . . . . . . . . 55<br>    Heterotrophic nutrition . . . . . . . . . . . . . . . . . 55<br>    Saprophytes and parasites . . . . . . . . . . . . . 66 | Unit 2 | Component 2 | Component 3 (Requirements for life) |

## Questions & Answers

# ■ Getting the most from this book

Exam-style questions

Commentary on the questions

Tips on what you need to do to gain full marks, indicated by the icon **e**

Sample student answers

Practise the questions, then look at the student answers that follow.

Commentary on sample student answers

Find out how many marks each answer would be awarded in the exam and then read the comments (preceded by the icon **e**) following each student answer. Annotations that link back to points made in the student answers show exactly how and where marks are gained or lost.

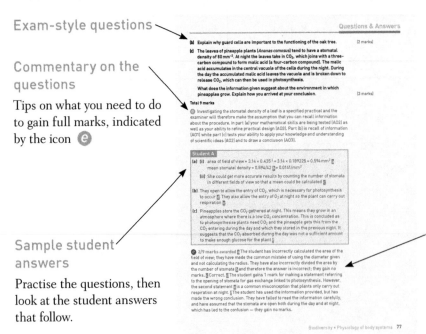

# ■ About this book

This guide will help you to prepare for the WJEC AS/A-level Biology Unit 2 exam. Your understanding of some of the principles in Unit 1 may be re-examined here too.

The **Content Guidance** covers all the concepts that you need to understand and facts that you need to know for the Unit 2 exam. It also includes *exam tips* and *knowledge checks* to help you to prepare for Unit 2. There are four aspects to the unit, reflected in this guide, which you will find useful during your revision:

1. Body systems have evolved, by a process of natural selection, from common ancestors, so there is an **evolutionary theme** running through Unit 2. It is important that you keep this in mind as you use the guide because it will help you to understand the concepts involved.

2. The unit also follows a **comparative approach**, which requires the understanding of a particular concept, for example, gas exchange, and how this concept relates to adaptations such as gills in fish, lungs in mammals and a tracheal system in insects. In each topic, the concepts are presented first and then the details of the adaptations of the different organisms are explained. It is a good idea to get your mind around these key ideas before you try to learn all the associated facts.

3. The **order** in which topics appear in the guide follows the order of the specification, with two exceptions. In topic 1, biodiversity comes before classification, as understanding of the evolution of biodiversity facilitates an understanding of classification. In topic 3, plant transport comes before animal transport, as this topic follows directly on from gas exchange in plants. Gas exchange and transpiration are clearly linked and there is scope for examiners to write questions on plants that incorporate both topics.

4. Included at the end of each topic is a brief description of the **practical work** that you should have undertaken during the course. In the majority of cases this provides opportunities for examiners to assess your **mathematical skills** as well as your practical skills. With the exception of the topic on biodiversity and classification, all of the other topics include the examination of microscope slides of various tissues and organs. Examiners may use photomicrographs or drawings of these tissues and organs and ask questions that relate the visible structures to their functions. You may also be asked to calculate the actual size of structures within the image, or to calculate the magnification of the image; both of which you should have done during Unit 1.

The **Questions & Answers** section will help you to:

- ■ familiarise yourself with many of the different question styles you can expect in the unit test
- ■ understand what the examiners mean by terms like 'describe' and 'explain'
- ■ interpret the question material — especially any data that the examiners give you
- ■ write concise answers to the questions that the examiners set

Each question in this section is attempted by two students, student A and student B. Their answers, along with the comments, should help you to see what you need to do to score a good mark — and how you can easily not score a good mark even though you probably understand the biology.

# Content Guidance

## ■ The evolutionary relationships between organisms

### Biodiversity

It is estimated that there are between 10 million and 100 million different species on the planet. In general, the greatest biodiversity is found in the tropics and this biodiversity decreases as one moves towards the poles. For example, coral reefs and tropical rainforests are the most diverse habitats on the planet.

Evolutionary history shows that the majority of all organisms are now extinct. The fossil record indicates that there have been five mass extinction events, when most species became extinct; these are referred to as **bottlenecks** in biodiversity. The fossil record shows that since the last of these events biodiversity has increased and new species have radiated out from pre-existing species.

### Assessing biodiversity

Biodiversity can be assessed in different ways:
- in a habitat
- within a species at both genetic and molecular levels

### Biodiversity of habitats

Sand dunes often have large numbers of marram grass growing on them and very few other plant species; they have low biodiversity. Tropical rainforests, in comparison, contain large numbers of different plant species; they have high biodiversity.

To investigate the biodiversity of a habitat, ecologists need to count the number of species present (species richness) and the number of individuals within each species population. Once the data has been collected it is possible to calculate the diversity of a habitat. Table 1, for example, shows data on different species found in two habitats, riffles and pools, within a stream.

| Species | Number in riffle | Number in pool |
|---|---|---|
| Minnow | 12 | 4 |
| Stonefly larva | 16 | 2 |
| Mayfly larva | 19 | 6 |
| Dragonfly larva | 9 | 1 |
| Freshwater shrimp | 22 | 14 |
| Bloodworm | 0 | 53 |

**Table 1** Biodiversity data for a stream

### Knowledge check 1

Which habitat in Table 1 has the highest species richness? Explain your answer.

Although the riffle appears to be more species diverse, it is important to be able to express diversity in numerical terms. This can be done by using the index of diversity, such as Simpson's diversity index, which is represented by the following formula:

$$D = 1 - \frac{\Sigma n(n-1)}{N(N-1)}$$

where $D$ is the diversity, $N$ is the total number of organisms of all species, $n$ is the total number of organisms of a particular species and $\Sigma$ is 'the sum of'.

Table 2 can be used to help calculate the diversity index of the pool.

| Species | Number ($n$) | ($n-1$) | $n(n-1)$ |
|---|---|---|---|
| Minnow | 4 | 3 | 12 |
| Stonefly larva | 2 | 1 | 2 |
| Mayfly larva | 6 | 5 | 30 |
| Dragonfly larva | 1 | 0 | 0 |
| Freshwater shrimp | 14 | 13 | 182 |
| Bloodworm | 53 | 52 | 2756 |
| **Total ($N$)** | 80 | $\Sigma$ | 2982 |
| $N-1$ | 79 | | |
| $N(N-1)$ | 6320 | | |

$D = 1 - (2982/6320)$
$D = 1 - 0.47$
$D = 0.53$

**Table 2** Calculation of the Simpson's diversity index for the pool

Any value calculated using Simpson's diversity index ranges between 0 and 1. The greater the value, the greater the sample diversity. When the diversity of the riffle is calculated, $D = 0.79$, therefore we can conclude that even though the species richness of the pool (six species present) is higher than that of the riffle (five species present) the biodiversity of the riffle is higher than that of the pool.

## Biodiversity within a species

Members of the same species all have the same genes, but different combinations of alleles. The number of different alleles present within the gene pool of a species is called the **genetic diversity** of the species. Polymorphism is the word used to describe the presence of several different forms or types of individuals among the members of a single species. For example, there are two main forms of the peppered moth (*Biston betularia*), which have different wing colours.

Polymorphism results from the presence of polymorphic genes. The shells of the land snail (*Cepaea nemoralis*), for example, may demonstrate a variety of distinct colours, including yellow, pink and brown. The shells may also be marked with one to five dark bands that have varying amounts of pigmentation. All of these characteristics are genetically determined.

**Knowledge check 2**

Use the data in Table 1 to confirm that the diversity of the riffle is 0.79.

The gene pool is all of the different alleles, of all of the different genes, in a population.

Polymorphic genes have more than one allele at each gene locus.

- Shell colour is controlled by six alleles that are at the same gene locus.
- Pigmentation of the bands is controlled by four alleles that are at a different gene locus.

Genetic variation within a species is commonly measured as the proportion of polymorphic gene loci across the genome. It is not practically feasible to count every single allele in a population, so researchers collect samples of DNA and analyse the base sequences to look for variations between individuals. The greater the variation in the DNA base sequence, the greater the genetic diversity of the species.

The DNA base sequences of elephant seal populations have been studied to assess their genetic diversity. The northern elephant seal (*Mirounga angustirostris*) population demonstrates reduced genetic variation, probably because of a population bottleneck: hunting reduced their population size to about 20 individuals at the end of the nineteenth century. The population of the northern elephant seal has since increased to over 30 000, but they have much less genetic variation than a population of southern elephant seals (*Mirounga leonina*) that was not so intensely hunted.

Genetic diversity is important because it is the basis of natural selection and survival of a species. A species with a high genetic diversity is likely to have some individuals with the characteristics that are required to cope with a change in the environment, so that some members of the species will survive. Low genetic diversity means that a species is more likely to become extinct if there are adverse environmental changes.

# Evolution by natural selection

The variety of living organisms that exists today has evolved as a result of natural selection. The theory of natural selection was proposed by Charles Darwin and Alfred Russel Wallace in the mid-nineteenth century.

Darwin studied the finches of the Galapagos Islands, a small group of volcanic islands off the coast of Ecuador. Darwin observed how the beaks of individual finches differed on the different islands, and also that the size and shape of the beak allowed the finches to exploit the particular source of food available on that island (see Figure 1). The general resemblance of the island finches to those on the mainland led Darwin to suggest that an ancestral population of finches must have colonised the different islands, shortly after their formation. Darwin suggested that the finches had developed from this **common ancestor**, and that each type of island finch had, over time, developed a type of beak that was adapted to exploit a particular food source. This is a classic example of **adaptive radiation**.

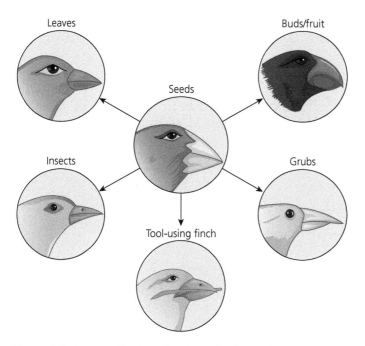

**Figure 1** Galapagos finches showing adaptive radiation

Following his 5-year voyage on the *Beagle*, Darwin started to develop his views on the mechanism by which these changes occurred. Darwin proposed that species evolve by a process of **natural selection**, which can be summarised by a series of observations and conclusions.

1. **Overproduction of offspring**: all organisms reproduce to give far more offspring than are needed simply to replace the parents.

2. **Constancy of number**: populations of organisms fluctuate, but population numbers remain fairly constant.

3. **Struggle for existence**: from these two observations, Darwin concluded that there must be a 'struggle for existence' between individuals of the same species, and that some individuals will not survive long enough to reproduce.

4. **Variation among offspring**: the offspring of any species that reproduces sexually show individual variations.

5. **Survival of the fittest**: Darwin concluded that those individuals that are best adapted to their environment will be more likely to survive in the 'struggle for existence'.

6. **Offspring resemble their parents**: those individuals that survive to reproduce are likely to produce offspring that are similar to themselves, therefore favourable adaptations are transmitted from one generation to the next.

7. **Formation of new species**: Darwin concluded that over successive generations, the characteristics of a population will slowly change. Those individuals that lack favourable characteristics will be less likely to survive and reproduce, and so their numbers will decline. Those individuals that exhibit favourable characteristics will survive and reproduce, and their numbers will increase.

The Galapagos finches example focuses specifically on the anatomical adaptations of the birds to the different environments, that is, the size and shape of their beaks. Natural selection will, however, lead to both physiological and behavioural adaptations to enable each species to become adapted to a particular environment. For example, the tree finches feed primarily on insects, so their digestive systems would also have had to adapt to a higher protein diet, when compared with their seed-eating ancestors. The woodpecker finch has evolved various behavioural adaptations that allow it to feed on insect larvae. This species is one of the few birds in the world that we have observed using 'tools'; it uses a small twig, or cactus spine, to pry the larvae out of small holes in cactuses, or from beneath the layer of bark.

# Classification

Classification places organisms into discrete groups with other closely related species. Classification systems are **hierarchical**, that is, larger groups are subdivided into smaller groups. Each group, or taxon, may contain a number of groups (taxa) lower in the hierarchy. Each group has features that are unique to that group — for example, all mammals feed their young on milk.

**Exam tip**

You will need to remember the names of the seven taxonomic groups and the order in which they are arranged. The use of a mnemonic can help you to remember the names and the order, for example, **K**ing **P**hilip **C**rossed **O**ceans **F**or **G**old and **S**ilver.

| Taxon | Human | Grey wolf |
|---|---|---|
| Kingdom | Animalia | Animalia |
| Phylum | Chordata | Chordata |
| Class | Mammalia | Mammalia |
| Order | Primata | Carnivora |
| Family | Hominidae | Canidae |
| Genus | *Homo* | *Canis* |
| Species | *sapiens* | *lupus* |

Increase in similarity of organisms in each taxon (group) ↓

**Table 3** The classification of humans and the grey wolf

Table 3 shows that as organisms are classified down the hierarchy they have more characteristics in common. Both humans and wolves have a backbone and feed their young on milk. The possession of a backbone is a characteristic that they share with birds, reptiles, amphibians and fish, which all belong to the **phylum** Chordata. However, feeding the young on milk is a characteristic that is unique to the **class** of mammals.

A species is a group of organisms that have similar characteristics and that can **interbreed** and produce **fertile** offspring.

The biological name of an organism follows a **binomial** (two-name) system that uses the name of its **genus** and species. The biological name is in Latin, which is a universal language and internationally recognised, and is printed in *italics*.

**Knowledge check 3**

What are the scientific names for humans and the grey wolf?

# Phylogenetic trees

Biological classification of organisms is based on the *presence of shared characteristics* and on the *evolutionary relationships* between organisms; this is known as a **phylogenetic** system. A phylogenetic tree represents the evolutionary history of different species and indicates common ancestors and lines of descent (Figure 2). Molecular data and the presence of homologous structures are used to help construct phylogenetic trees.

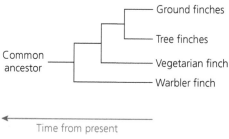

**Figure 2** Phylogenetic tree showing the evolutionary relationships between finches found on the Galapagos Islands

**Knowledge check 4**

Which two groups of finches in Figure 2 are the most closely related?

Organs in one species that have a similar basic structure to organs in other species are referred to as being **homologous structures**; for example, the pentadactyl limb in vertebrates (Figure 3). The forelimbs of a crocodile and the wings of a bird may differ in function, but they are related to each other through common descent.

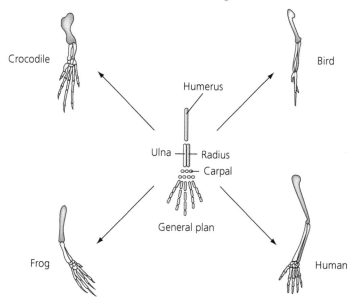

**Figure 3** Pentadactyl limb

**Analogous structures** have the same function but are not derived from the same organ in a common ancestor; for example, the wings of an insect and the wings of a bird. Analogous structures arise due to a process of **convergent evolution**, as a result of adaptation to similar environmental conditions. Convergent evolution may make it difficult to distinguish between homologous structures and analogous structures, and has resulted in organisms being incorrectly grouped together.

**Biochemical analysis** has allowed more accurate classification to confirm evolutionary relationships. A mutation results in a change in the base sequence of DNA, and as species diverge from a common ancestor, each distinct lineage accumulates changes in its DNA base sequences. DNA base sequence comparisons provide a reliable indicator of similarity without the problem of morphological convergence.

Changes in the DNA base sequence will alter the amino acid sequence in a protein. Comparisons of the amino acid sequences of common proteins can also be used to estimate relatedness between species. However, as the number of common proteins is limited, most new studies use RNA and DNA comparisons.

Homologous structures and biochemical analysis are both used to work out evolutionary relationships between organisms. Biologists generally accept that all organisms have evolved from a **common ancestor** via adaptive radiation:

- Species that have many characteristics in common are closely related and share a recent common ancestor in their evolutionary history.
- Species that have fewer characteristics in common are not closely related and share a less recent common ancestor in their evolutionary history.

## Schemes of classification

Schemes of **classification** are based upon the best evidence available and are subject to change. Early biologists recognised two kingdoms — animals and plants. The invention of the microscope in the 1880s led to the discovery of unicellular organisms and the Protoctista and Prokaryotae kingdoms were suggested. The current five-kingdom system, proposed in the 1960s, was based on cell type, organisation and nutrition.

| Kingdom | Main characteristics of organisms | | |
| --- | --- | --- | --- |
| | Cell type | Organisation | Nutrition |
| Animalia | Eukaryotic cells that lack a cell wall | Multicellular and possess a nervous system | Heterotrophic |
| Plantae | Eukaryotic cells that possess chloroplasts and a cell wall made of cellulose | Multicellular | Autotrophic |
| Fungi | Eukaryotic cells that possess a cell wall made of chitin | Mainly multicellular, have a body made of hyphae forming a mycelium, and reproduction is via spores | Heterotrophic (either saprophytes or parasites) |
| Protoctista | Eukaryotic cells | Mostly unicellular but can be multicellular | Autotrophic or heterotrophic |
| Prokaryotae | Prokaryotic cells that possess a cell wall made of murein/peptidoglycan | Unicellular | Autotrophic or heterotrophic |

**Table 4** The general characteristics of the five kingdoms

### Exam tip

Due to the majority of the Protoctista being unicellular it is easy to confuse organisms within this kingdom with those of the Prokaryotae. If an exam question asks you to identify which kingdom a unicellular organism belongs to look for the presence or absence of membrane-bound organelles.

### Knowledge check 5

The dorsal fins of the shark and the dolphin are evidence of common ancestry. True or false?

### Exam tip

If you are asked about the characteristics of organisms in different kingdoms, remember your knowledge of cell structure, from Unit 1, as this will provide most of the answers.

### Knowledge check 6

The composition of the cell walls in plants, fungi and prokaryotes is different. Name the component that makes up the cell walls in these three kingdoms.

# Three domains and five kingdoms

Recent biochemical evidence has shown that members of the prokaryotic kingdom have some fundamental differences. Some microorganisms, known as **extremophiles**, exist in a wide variety of environmental conditions, including extremes of temperature, pH, salinity and pressure. Although they have prokaryotic cells, the biochemistry of their cells is distinctly different to that of other forms of life, in particular with respect to the sequence of nucleotide bases in the ribosomal RNA. These differences have led to the development of a new scheme of classification, which suggests that all organisms have evolved along three distinct lineages, which are called **domains**. The organisms of each domain share a unique and distinctive pattern of ribosomal RNA, which establishes their close evolutionary relationship. The three domains of life are:

- **Bacteria** (or Eubacteria), which are the true bacteria.
- **Archaea** (or Archaeabacteria), which are the extremophile prokaryotes.
- **Eukarya**, which includes all eukaryotic organisms: animals, plants, fungi and protoctists.

rRNA sequencing results suggest that bacteria diverged first, as the archaea and eukarya are more closely related to each other than either is to the bacteria (Figure 4).

**Five kingdoms**

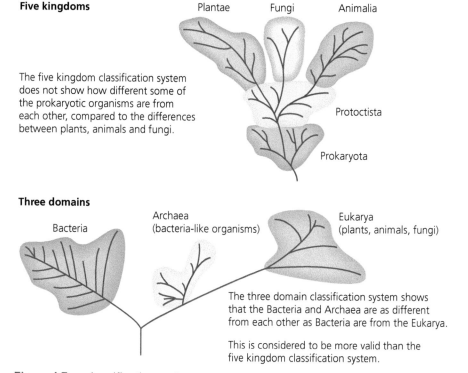

The five kingdom classification system does not show how different some of the prokaryotic organisms are from each other, compared to the differences between plants, animals and fungi.

Plantae   Fungi   Animalia

Protoctista

Prokaryota

**Three domains**

Bacteria

Archaea (bacteria-like organisms)

Eukarya (plants, animals, fungi)

The three domain classification system shows that the Bacteria and Archaea are as different from each other as Bacteria are from the Eukarya.

This is considered to be more valid than the five kingdom classification system.

**Figure 4** Two classification systems

## Practical work

During this topic you should have completed some form of practical activity to investigate biodiversity. There are many different techniques that could have been carried out, using quadrats, transects, sweep nets or kick sampling in a stream. Regardless of the technique used, it is impossible to count all of the individuals in a population, therefore you would have taken samples. There are two important principles that must be taken into account when sampling populations:

■ The sample must be large enough so as to be **representative** of the population as a whole.
■ The sample should be taken at **random** so that the data are **unbiased**. If the sample is not taken at random then the data obtained are unlikely to be representative of the true population.

For example, if you were assessing the biodiversity of plants in a meadow, you would need to include the following steps:

■ Mark out a grid using two tape measures at right angles to each other.
■ Choose pairs of numbers at **random** (for example, using a random-number table) to generate coordinates.
■ Place the quadrat at the coordinate and record the species present within the quadrat.
■ **Repeat** the process a number of times to ensure that the sample is large enough to be representative.

## Summary

After studying this topic you should be able to demonstrate and apply your knowledge and understanding of:

■ the term biodiversity and that many factors can cause biodiversity to vary both spatially and over time
■ the assessment of biodiversity:
 — in a habitat, including the use of Simpson's diversity index to calculate biodiversity
 — within a species, by looking at the variety of alleles in the gene pool of a population
 — at a molecular level, using DNA fingerprinting and sequencing
■ natural selection in generating biodiversity
■ the anatomical, physiological and behavioural adaptations of organisms to their environment

■ the classification of organisms, which places organisms into discrete and hierarchical groups with other closely related species, and is based on their evolutionary relationships
■ how the relatedness of organisms can be assessed by comparing both morphology and biochemistry
■ the concept of species and the use of the binomial system in naming organisms
■ the tentative nature of classification, and the ability to compare the three domain classification system with the five kingdom classification system
■ the characteristic features of the kingdoms Prokaryotae, Protoctista, Plantae, Fungi and Animalia

# ◼Adaptations for gas exchange

## Gas exchange surfaces

Living organisms exchange gases between their cells and the environment via **diffusion**. The *rate* of diffusion of gases is increased by:

- increasing the surface area
- reducing the length of the diffusion pathway
- increasing the steepness of the concentration gradient

For efficient exchange all gas exchange surfaces must have the following properties:

- be **permeable** to gases
- be **moist,** as gases must dissolve before they can diffuse across membranes
- have a **large surface area** over which the gases can diffuse
- be **thin,** so as to provide a short diffusion pathway

Large active animals can also have **ventilating mechanisms** to maintain steep concentration gradients across the gas exchange surface. Because gas exchange surfaces are moist, water will evaporate from them if they are in contact with air. To overcome this problem, terrestrial animals have evolved *internal* gas exchange surfaces to help reduce water loss.

**Knowledge check 7**

State three features that are common to *all* gas exchange surfaces.

## Surface area-to-volume ratio

All organisms require $O_2$ for cellular respiration. The *quantity* of $O_2$ required by an organism is proportional to its volume. The *rate* of uptake of $O_2$ from the environment is proportional to its surface area. Figure 5 shows the relationship between an increase in size and the surface area-to-volume ratio.

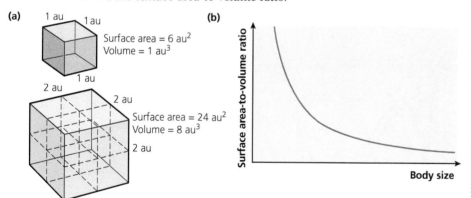

**(a)**
1 au   1 au

Surface area = 6 au²
Volume = 1 au³

1 au
2 au

2 au

Surface area = 24 au²
Volume = 8 au³

2 au

**(b)**
Surface area-to-volume ratio

Body size

**Figure 5** (a) The effect of increasing size on the surface area and volume of a cube (au = arbitrary units); (b) the relationship between size and surface area-to-volume ratio

The smaller (pink) cube has a surface area-to-volume ratio of 6:1, whereas the larger (blue) cube has a surface area-to-volume ratio of only 3:1. This demonstrates that as organisms get bigger their surface area-to-volume ratio decreases. This is also shown by the curve in Figure 5(b).

A small unicellular organism, such as an amoeba, has a very large surface area-to-volume ratio and short diffusion pathways to all parts of the cell (Figure 6).

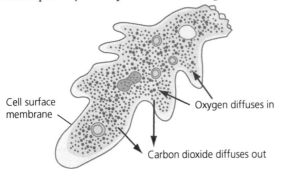

**Figure 6** Amoeba

As organisms get larger, their demand for oxygen increases but their surface area-to-volume ratio decreases. Multicellular organisms have therefore evolved adaptations to maintain adequate gas exchange so that they can meet this increased demand for $O_2$. For example, the flattened body of a flatworm both increases its surface area-to-volume ratio and reduces the diffusion pathway (allowing its body surface to continue to act as the gas exchange surface). Arthropods and vertebrates have evolved specialised gas exchange surfaces.

# Gas exchange in mammals

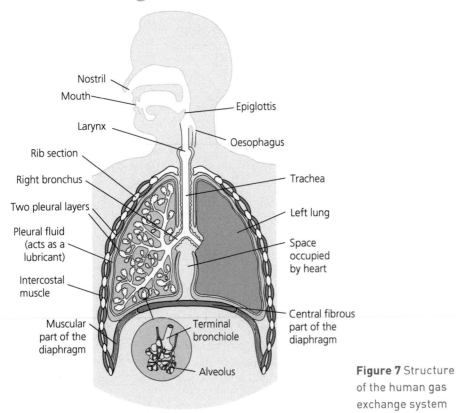

**Figure 7** Structure of the human gas exchange system

Figure 7 shows the generalised structure of the human gas exchange system. The trachea, bronchi and bronchioles contain rings of cartilage, which prevent these airways from collapsing under the negative pressure produced in the lungs during inspiration. The ciliated epithelium lining the trachea also contains goblet cells, which secrete mucus to trap and remove dust particles and microorganisms that may be present in inspired air. The cilia beat in a wave-like manner, moving the mucus upwards and out of the lungs.

## Ventilation

**Ventilation** in mammals is brought about by changing the pressure inside the lungs; air always moves down a pressure gradient. Inspiration is an active process that is brought about by the contraction of muscles (Figure 8(a)). Expiration is a passive process that is caused by the recoil of elastic tissues (Figure 8(b)).

Table 5 summarises the events that occur during inspiration and expiration in a mammal.

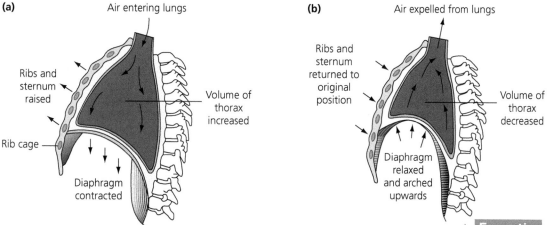

**Figure 8** Mechanism of breathing showing (a) inspiration and (b) expiration

|  | Inspiration | Expiration |
|---|---|---|
| Intercostal muscles | Contract | Relax |
| Rib cage | Moves upwards and outwards | Moves downwards and inwards |
| Diaphragm muscles | Contract | Relax |
| Diaphragm | Flattened | Dome shaped |
| Volume of thoracic cavity | Increases | Decreases |
| Pressure in thoracic cavity and lungs | Decreases (below atmospheric pressure) | Increases (above atmospheric pressure) |
| Direction of air flow | Into the lungs | Out of the lungs |

**Table 5**

Gas exchange takes place in the alveoli. Figure 9 shows the structure of an alveolus and associated capillary. The arrows show $O_2$ diffusing from the alveolus into the blood and $CO_2$ diffusing from the blood into the alveolus. Water continually evaporates from the lining of the alveolus; the water vapour is breathed out during expiration. Table 6 shows how the lungs are adapted for gas exchange.

**Knowledge check 8**

State two advantages to a mammal of having internal lungs.

**Ventilation** is the movement of the respiratory medium over the respiratory surface.

**Exam tip**

You might be asked to describe the process of inspiration in a mammal. Make sure that your answer is well structured and clearly links cause and effect. You will not be rewarded for describing all the correct components in the wrong order!

**Knowledge check 9**

Explain the advantage of ventilation to a large active organism.

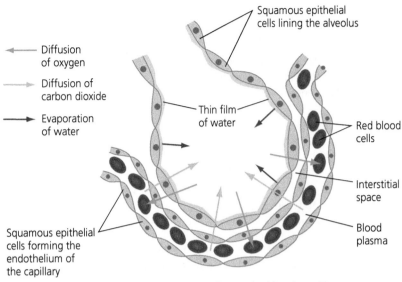

Squamous epithelial cells lining the alveolus

← Diffusion of oxygen

→ Diffusion of carbon dioxide

→ Evaporation of water

Thin film of water

Red blood cells

Interstitial space

Blood plasma

Squamous epithelial cells forming the endothelium of the capillary

**Figure 9** Gas exchange between an alveolus and a blood capillary

| Feature | Adaptation |
|---|---|
| Moist surface | ■ Tissue fluid lining the alveolus allows gases to dissolve and diffuse across |
| Large surface area | ■ Millions of alveoli<br>■ An extensive capillary network surrounding each alveolus |
| Short diffusion pathway | ■ The alveolar wall is a single layer of flattened epithelial cells<br>■ The capillary wall is a single layer of flattened endothelial cells |
| Maintenance of steep concentration gradients | ■ Ventilation ensures that the $O_2$ concentration in the alveolus is high<br>■ Dense capillary network and blood flow ensure that the $O_2$ concentration in blood entering the alveolar capillaries is low |

**Table 6** Adaptations shown by mammalian lungs for gas exchange

## Knowledge check 10

Figure 10 is a photomicrograph of a section through the mammalian lung. Describe and explain how three features shown in the micrograph are adaptations for efficient gas exchange.

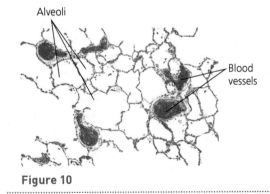

Alveoli

Blood vessels

**Figure 10**

# Gas exchange in other terrestrial animals

Air is the respiratory medium used by terrestrial animals. As gas exchange surfaces must be moist, terrestrial animals face dehydration due to water evaporating from respiratory surfaces.

## Annelids (earthworms)

Earthworms use their body surface for gas exchange. They secrete mucus to ensure that their 'skin' remains moist. Annelids are restricted to damp areas, which helps to prevent desiccation. They are elongated, which provides a large surface area-to-volume ratio. They have a closed circulatory system, with a well-developed capillary network close to the body surface, which provides a short diffusion pathway. Earthworm blood contains haemoglobin that has a high affinity for oxygen. Oxygen is transported round the body, which helps to overcome the increase in diffusion distance.

## Insects

Insects have a waterproof exoskeleton that is made of chitin. Openings called **spiracles** on the exoskeleton lead to a branched, chitin-lined system of tracheae. The gas exchange surfaces are the **tracheoles**, which come into contact with every tissue. The advantages are that every tissue is supplied directly with oxygen and no circulation or haemoglobin is needed. Large insects contract and relax muscles in their thorax and abdomen, causing rhythmical movements that ventilate the tracheoles, maintaining a concentration gradient. The tracheal system of an insect is shown in Figure 11.

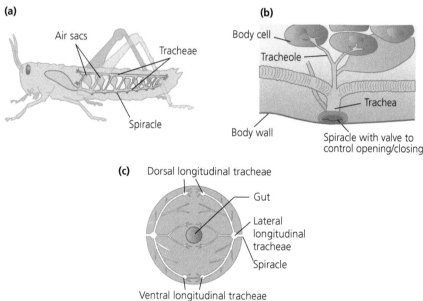

**Figure 11** (a) The tracheal system of an insect; (b) the relationship between spiracles, tracheae and tracheoles; (c) transverse section of the insect tracheal system

# Vertebrates

Vertebrates are large animals and the surface area of their skin is insufficient to act as a gas exchange surface. Terrestrial vertebrates have therefore evolved **internal lungs** for gas exchange. These provide a large surface area and minimise water loss as they are within the body cavity.

## Amphibians

For example, a frog:

- The larval form (tadpole) is aquatic and uses gills for gas exchange.
- The adult frog uses its moist skin for gas exchange when it is inactive, but when active it uses its lungs for gas exchange. The adult frog's lungs are a pair of thin-walled sacs.

## Reptiles

- Reptile lungs are more efficient than amphibian lungs as they are more highly folded, giving them a much greater surface area.
- Reptiles also have a ribcage and can therefore ventilate their lungs.

## Birds

- The ventilation of bird lungs is similar to that of reptiles, but in birds the effectiveness is increased by the presence of air sacs.
- No gas exchange occurs in the air sacs, but their arrangement increases the efficiency of lung ventilation, by acting as bellows.

# Gas exchange in fish

**Water** is the respiratory medium used by fish. Fish have evolved gills (Figure 12) with **gill lamellae** that are adapted to gas exchange in the ways shown in Table 7. Water is a **dense medium** and passes over the gills in a **unidirectional flow**.

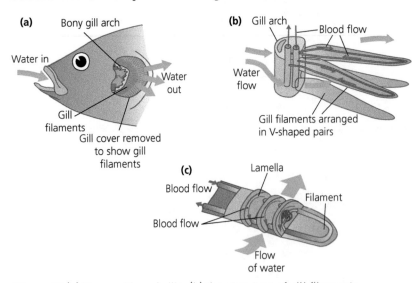

**Figure 12** (a) The position of gills; (b) the structure of gill filaments; (c) blood flow in lamellae

> **Knowledge check 11**
>
> Birds and mammals are endothermic ('warm blooded') and maintain a constant body temperature. Explain why endothermic animals must have well-developed gas exchange mechanisms.

| Feature | Adaptation |
|---------|-----------|
| Large surface area | ■ Millions of gill lamellae |
| Short diffusion pathway | ■ Gill lamellae are thin |
| Maintenance of steep concentration gradients | ■ Ventilation (oxygenated water is forced over the gills) ensures that the $O_2$ concentration in the water entering the gills is high <br> ■ Dense capillary network and blood flow ensure that the $O_2$ concentration in blood entering the gill lamellae is low <br> ■ Blood also contains haemoglobin with a high affinity for oxygen |

**Table 7** The adaptations shown by gill lamellae for gas exchange

# Maintaining steep concentration gradients

Some types of fish, for example sharks, must keep swimming to ensure that oxygenated water passes over the gills. Bony fish, such as groupers, have evolved ventilation mechanisms that can force water over the gills even when they are stationary in the water.

# Ventilation

The gills of a fish are located between the buccal cavity and the opercular cavity (Figure 13). Ventilation of the gills is brought about by pressure changes in these two cavities.

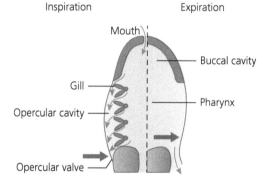

**Figure 13** Longitudinal section of fish head showing the buccal and opercular cavities during inspiration and expiration

## Inspiration

During inspiration the fish opens its mouth and lowers the floor of the buccal cavity. This increases the volume of the buccal cavity and decreases its pressure, causing water to enter through the mouth. At the same time the volume of the opercular cavity is increased so that its pressure decreases below that of the buccal cavity. This causes water to be drawn across the gills down a pressure gradient. (The operculum valve is closed to prevent the entry of water through the operculum.)

## Expiration

During expiration the fish closes its mouth and raises the floor of the buccal cavity. This decreases the volume of the buccal cavity and increases its pressure, causing water to be forced over the gills. As water enters the opercular cavity, the resulting increase in pressure causes the operculum valve to open, and water exits the fish.

Figure 14 shows the pressure changes that occur in the buccal and opercular cavities during inspiration and expiration. For the majority of the ventilation cycle water is being forced over the gills as the pressure in the buccal cavity is greater than the pressure in the opercular cavity.

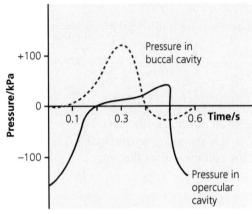

**Figure 14** Fish ventilation cycle

> **Exam tip**
>
> Ventilation in fish is more complicated than ventilation in mammals, however the principles are the same. The respiratory medium, air or water, moves down a pressure gradient. Changes in pressure within a cavity, such as in the lungs or in the opercular cavity, are brought about by changes in volume, which are generally caused by the contraction of muscles.

## Countercurrent mechanism

The concentration of (dissolved) oxygen in water is much lower than in air and therefore bony fish have evolved an efficient mechanism of gas exchange. Water flows over the gills in the opposite direction to that of the blood in the gill lamellae — this is known as a **countercurrent flow** (see Figures 12(c) and 15).

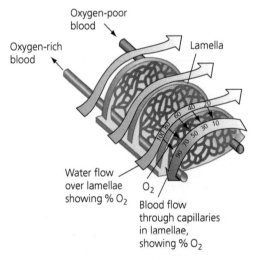

**Figure 15** A single gill lamella

The countercurrent mechanism makes gas exchange very efficient because the concentration of $O_2$ in the water is always higher than the concentration of $O_2$ in the blood, so that equilibrium is never reached (see Figure 16). This enables $O_2$ to diffuse into blood along the whole length of the gill lamellae. The countercurrent mechanism allows bony fish, such as herring, to extract about 80% of oxygen in water.

In cartilaginous fish (sharks and rays), such as the dogfish, water flows over the gills in the same direction as that of the blood in the gill lamellae — this is known as **parallel flow**. This is a less efficient mechanism for gas exchange because the concentration of $O_2$ in the water and the blood reaches equilibrium part way across the gill lamellae. As a result the dogfish can extract only about 50% of the oxygen in water.

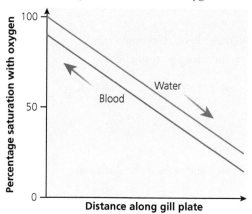

**Figure 16** Oxygen concentrations in water and blood

**Knowledge check 12**

Name the gas exchange surface that is found in each of the following organisms: an amoeba, earthworm, insect, fish and mammal.

# Gas exchange in plants

Plants require $CO_2$, as well as light, for photosynthesis. Their leaves are adapted for both light absorption and gas exchange. Plants rely entirely on **diffusion** for the exchange of gases (Figure 17).

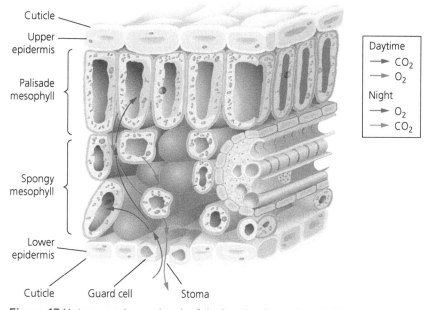

**Figure 17** Net gas exchange in a leaf during the day and at night

# Adaptations for light absorption

- Plants can orientate their leaves towards the light to expose a greater surface area.
- Leaves are flat and have a large surface area to absorb maximum light.
- Leaves are thin to allow light to penetrate to lower tissue layers.
- The cuticle and epidermis are transparent to allow light to penetrate mesophyll tissue.
- Palisade cells are elongated to reduce the total number of cell walls that would absorb light (preventing it from reaching the chloroplasts).
- Palisade cells contain many **chloroplasts** to maximise light absorption.
- Chloroplasts move inside cells to gain the best position for absorbing light.

# Adaptations for gas exchange/$CO_2$ absorption

- Leaves are thin to reduce the diffusion distance.
- The **spongy mesophyll**:
  - has a large surface area for gas exchange
  - contains air spaces to allow the circulation of gases and reduce the diffusion pathway of $CO_2$ into cells
  - is moist for the absorption of $CO_2$
- **Stomatal pores** allow the entry of gases into the air spaces.

# Adaptations to reduce water loss

Plants face the same problems as terrestrial animals — gas exchange surfaces must be moist in order for gas exchange to occur. As a consequence, however, water evaporates from the gas exchange surface and dehydrates the organism. Plants (like animals) cannot prevent water loss but they have adaptations to **reduce** it:

- Leaves have a **waxy cuticle** on the **upper epidermis** to reduce water loss by evaporation (however, this prevents gaseous exchange).
- The majority of water loss from a plant is via the stomatal pores. The stomatal pores are located on the **lower epidermis**, which helps to reduce water loss via evaporation.
- **Guard cells** surrounding the stomata can change shape:
  - they open the stomatal pores during the day to allow the entry of $CO_2$ (i.e. gas exchange) for photosynthesis
  - they close the stomatal pores at night to reduce water loss

## Mechanism of stomatal opening

Guard cells are the only cells in the lower epidermis to contain chloroplasts. During the day (when light is available) photosynthesis occurs, which results in:

- increased production of **ATP**
- a lower $CO_2$ concentration in the guard cells

The ATP is used to **actively transport** $K^+$ into the guard cells (from the epidermal cells). The lower $CO_2$ concentration triggers the conversion of insoluble starch to soluble malate. The $K^+$ and the malate lower the **water potential** in the guard cells. Water moves into the guard cells via osmosis. This causes the guard cells to swell and become turgid. The inner cell wall is thicker (and less elastic) than the outer cell wall, so as the cells swell they curve apart — opening the stomatal pore (as shown in Figure 18).

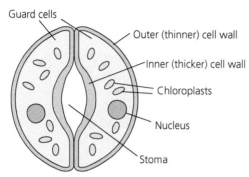

**Figure 18** Guard cells

## Practical work

The investigation of stomatal numbers in leaves is a relatively simple procedure and involves preparing an epidermal impression by coating the leaf surface with nail varnish. The dried layer of nail varnish is peeled off and placed on a microscope slide.

This practical allows examiners to assess your mathematical skills in calculating stomatal density and to apply this in different contexts, for example comparisons between the upper and lower epidermis or between two different species.

**Figure 19** Drawing from a microscope preparation of the lower epidermis of a leaf

Figure 19 shows a drawing from a photomicrograph of the lower epidermis of a leaf at 400× magnification. The diameter of the field of view is 435 μm. From this information we can calculate the stomatal density in mm², as follows.
- To convert from μm to mm the diameter must be divided by 1000 = 0.435 mm
- If the diameter is 0.435 mm then the radius is 0.2175 mm
- The area (of a circle) = $\pi r^2$, therefore the area of the field of view = $3.14 \times 0.2175^2 = 0.149$ mm² (to 3 significant figures)
- There are 13 stomata in the field of view, therefore the stomatal density = 13/0.149 = 87.2 stomata mm⁻²

➡

**Knowledge check 13**

Explain why it would be a disadvantage to a plant if the stomata remained permanently open.

**Exam tip**

Many students confuse the stomatal pore with the guard cell and refer to water moving into the stomata by osmosis. Some students also fail to appreciate that guard cells function in pairs, and as they curve apart from each other the stomatal pore opens.

**Exam tip**

As in this example, you are likely to be given the diameter of the field of view. Make sure that you remember to calculate the radius before calculating the density. You will also find it easier if you convert the radius into the units required for the answer.

During this topic you will have also made observations of tissues involved in gas exchange.

- A dissection of a fish head, which will have enabled you to better understand the mechanism of ventilation, and properties of the gas exchange system in fish.
- Examination of microscope slides of tissues from the mammalian lungs and trachea, as well as a dicotyledon leaf. In the exam you may be given photomicrographs or drawings of these tissues. It is important that you can identify the various structures and link them to their functions.

## Summary

After studying this topic you should be able to demonstrate and apply your knowledge and understanding of:

- the use of the body surface for gas exchange in small animals
- the need for specialised gas exchange surfaces as body size and metabolic rate increase, and that these gas exchanges surface are adapted to particular environmental conditions — for example gills in aquatic environments and lungs for terrestrial environments
- the features that are common to all gas exchange surfaces and the need for ventilating mechanisms in large active animals that have higher metabolic rates

- the human breathing system and how its structure relates to its function in ventilation and gas exchange
- the insect tracheal system as an adaptation to a terrestrial environment
- ventilation and gas exchange in bony fish, including a comparison of countercurrent flow with parallel flow in different fish species
- the role of the angiosperm leaf:
  - as an organ of gaseous exchange
  - including the role of different tissues in allowing the plant to photosynthesise effectively

# Adaptations for transport

## Transport in plants

Pre-existing knowledge

The uptake of mineral ions from the soil involves active transport, which you studied in detail in Unit 1. Remember that active transport requires specific carrier proteins in the plasma membrane and the use of ATP (produced during respiration in mitochondria) to transport ions against a concentration gradient.

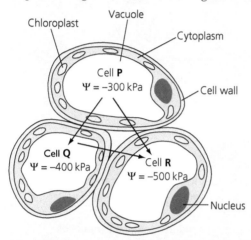

**Figure 20** Water movement (indicated by arrows) between three mesophyll cells in a leaf; cell P has the highest water potential and the water potential of cell Q is higher than that of cell R

The movement of water involves osmosis, which you also studied in Unit 1:

- Osmosis is the movement of water molecules down a water potential gradient through a selectively permeable membrane.
- The water potential of a cell can be lowered by increasing the concentration of solutes, such as mineral ions, in the cytoplasm.

In order to photosynthesise, the stomatal pores on a leaf must be open to allow gas exchange. However, this results in water loss from the plant and is linked to the movement of water through the xylem.

Plants are large, multicellular organisms and have well-developed vascular tissue to allow the transport of water, mineral ions and organic solutes. Water and mineral ions are transported from the roots to the leaves of plants in the **xylem**. Organic solutes, such as sucrose, are transported from the leaves to other parts of the plant in the **phloem**.

# Plant anatomy

Figure 21 shows the distribution of xylem and phloem tissue in a root and stem of a dicotyledonous plant. Table 8 summarises the functions of the different structures within a root.

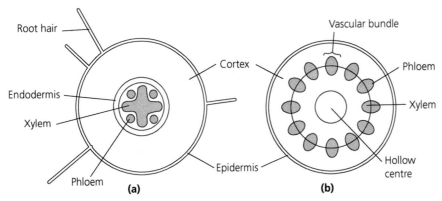

**Figure 21** The distribution of xylem and phloem in cross-sections of (a) a root and (b) a stem

| Structure | Function |
|---|---|
| Epidermis | Outer layer of cells, some of which are specialised into root hair cells — these provide an increased surface area for the uptake of ions and water |
| Root cortex | Is made of parenchyma cells, which provide mechanical support to the root |
| Endodermis | Layer of cells that surround the pericycle<br>Endodermal cells have a Casparian strip around them which is made of suberin (waxy substance) that waterproofs their cell walls |
| Pericycle | Contains the vascular tissue (xylem and phloem) |
| Xylem | Transports water and mineral ions within the plant |
| Phloem | Transports sucrose (and other organic solutes) within the plant |

**Table 8** The function of the different tissues within a root

Figure 22 is a photograph of the central region (the stele) of a dicotyledonous root, showing the distribution of xylem and phloem tissue as well as the endodermis.

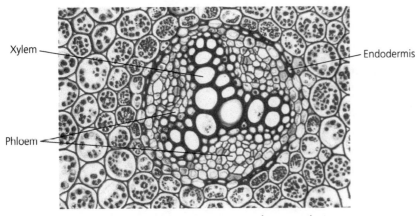

**Figure 22** Light micrograph of the central region (the stele) of a dicotyledonous root

The distribution of xylem and phloem is different in the root and in the stem (see Figure 21(b) and Figure 23). The vascular tissue is located within the central stele in the root, whereas it is located within discrete **vascular bundles** towards the periphery of the stem. The phloem is located in the outer part of the vascular bundle and the xylem is in the inner part.

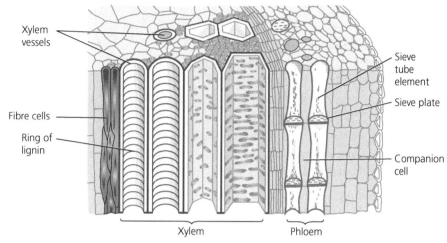

**Figure 23** Transport pathways in a plant system; xylem vessels and phloem sieve tubes in longitudinal and transverse sections

# Vascular tissue

## Structure of xylem

Xylem tissue is made up of four types of cell:
- vessels (elements)
- tracheids
- fibres
- parenchyma

**Vessels** and **tracheids** are composed of dead, elongated cells with pits in them (see Figure 24). Vessels are long, tubular structures that are formed by the end-to end fusion of vessel elements and the breakdown of their end walls. Their walls are waterproofed and strengthened by **lignin**. The function of vessels and tracheids is to transport water and mineral ions. The **fibres** and **parenchyma** provide support to the tissue.

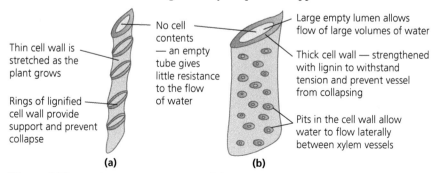

**Figure 24** Two xylem vessel elements (a) a narrow vessel thickened with rings; (b) a wider vessel with pits to allow lateral movement of water

**Exam tip**

Lignin provides strength, which is important to prevent the xylem vessels from **collapsing** due to the negative pressure inside the vessels during transpiration.

## Structure of phloem

Phloem tissue is made up of four types of cell:

- sieve tube elements
- companion cells
- fibres
- parenchyma

Sieve tubes are formed by the end-to-end fusion of **sieve tube elements**. They lack a nucleus, and the cytoplasm forms a thin layer around the periphery of the cell. The cell walls at the ends of each cell are perforated to form **sieve plates**, which allow the cytoplasm from one cell to run into adjacent cells. The function of sieve tubes is to transport organic solutes, for example, sucrose and amino acids.

**Companion cells** lie next to sieve tubes and are linked to them via **plasmodesmata**. Companion cells contain many cell organelles, especially **mitochondria**, and are involved in the loading and unloading of sieve tubes with solutes.

Figure 25 shows the structures of the sieve tube elements and the companion cells — note the sieve pores and the plasmodesmata, which adapt these cells to their specific functions. The **fibres** and **parenchyma** provide support.

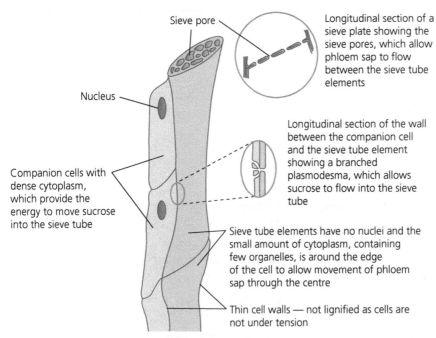

Sieve pore

Longitudinal section of a sieve plate showing the sieve pores, which allow phloem sap to flow between the sieve tube elements

Nucleus

Longitudinal section of the wall between the companion cell and the sieve tube element showing a branched plasmodesma, which allows sucrose to flow into the sieve tube

Companion cells with dense cytoplasm, which provide the energy to move sucrose into the sieve tube

Sieve tube elements have no nuclei and the small amount of cytoplasm, containing few organelles, is around the edge of the cell to allow movement of phloem sap through the centre

Thin cell walls — not lignified as cells are not under tension

**Figure 25** Phloem sieve tubes and companion cells

### Knowledge check 14

The conducting cells of xylem (vessels and tracheids) are dead when mature, yet the conducting cells of phloem (sieve tube elements) are alive. Explain the reason for this difference.

# The transport of water and mineral ions

## Uptake at the root

Most absorption of water is through the **root hairs**, which provide a large surface area for absorption. Ions are absorbed by **diffusion** and **active transport**, while water is absorbed by **osmosis** from a higher water potential in the soil water to a lower water potential in the xylem.

Water and mineral ions move through the root by three pathways (Figure 26):

- **apoplast pathway** — through the cell walls and the spaces between cells
- **symplast pathway** — through the cytoplasm via plasmodesmata
- **vacuolar pathway** — from vacuole to vacuole in adjacent cells

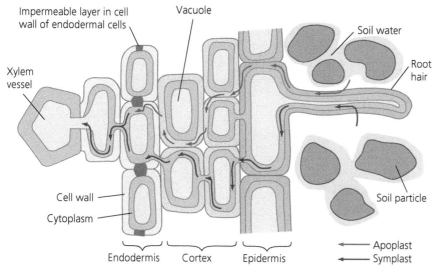

**Figure 26** The pathways taken by water as it moves from the soil, into a root hair, across the cortex, through the endodermis and into the xylem; plasmodesmata are shown as the green shaded areas passing through the cell walls, but they are not as big as shown here

### The endodermis

**Suberin** (a waterproof substance) is deposited in the cell walls and forms bands called **Casparian strips** in the endodermal cells, which block the apoplast pathway. Mineral ions are actively transported into the cytoplasm of the cells (symplast pathway). This lowers the water potential of the cells, causing water to move into the symplast pathway by osmosis. This helps to generate a water potential gradient across the root, drawing water in from the soil. The blocking of the apoplast pathway by the Casparian strips allows the endodermis to **selectively** uptake ions from the soil.

## Movement of water through the stem

There are three possible theories that explain the movement of water up the xylem:

- root pressure theory
- cohesion–tension theory
- capillarity/adhesion theory

**Knowledge check 15**

What is the difference between the apoplast and symplast pathways, and which one is blocked at the endodermis?

**Exam tip**

If plant roots are treated with a respiratory poison, such as cyanide, then the uptake of minerals will stop, as no ATP is generated.

## Root pressure theory

Endodermal cells actively transport ions into the xylem vessels. This lowers the water potential so water enters the xylem by osmosis, creating **hydrostatic pressure** (root pressure). This root pressure forces water up the stem, but it is not sufficient to push water to the leaves at the top of tall plants.

## Cohesion–tension theory

This involves the transpiration of water from the leaf. Water evaporates from the spongy mesophyll cells into the air spaces and diffuses out of the stomata down a water potential gradient. This sets up a water potential gradient across the leaf from a higher water potential in the xylem to a lower water potential in the air spaces.

Water molecules demonstrate a property known as **cohesion** — they are bonded together by weak hydrogen bonds. As water is drawn out from the top of the xylem, more water is pulled up the xylem to replace it. The pulling action of transpiration stretches the water column in the xylem so that it is under **tension**. In this way transpiration from the leaves pulls water up the stem in a continuous column, which is known as the **transpiration stream**. Figure 27 shows the events that take place during transpiration.

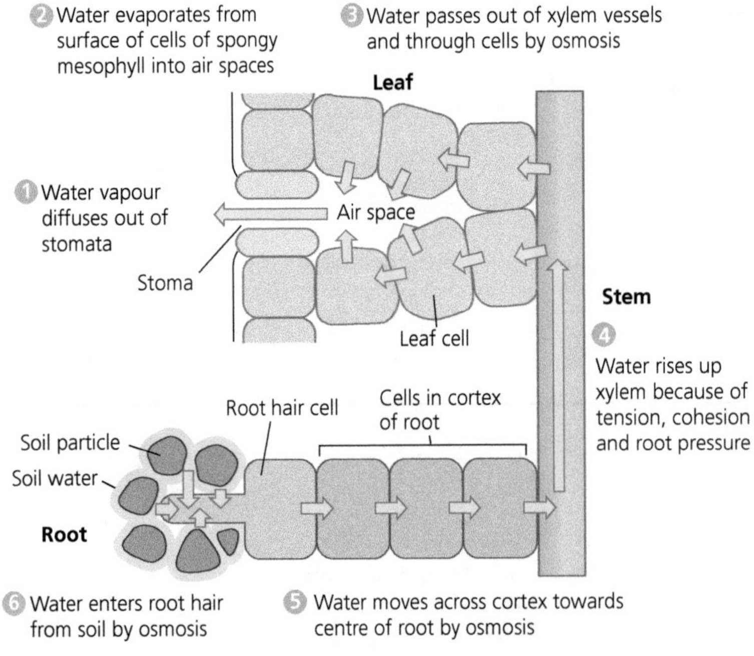

> Transpiration is the loss of water vapour from the leaves of a plant.

**Figure 27** Transpiration — the movement of water through a plant

## Adhesion theory

Xylem vessels are very narrow and have a hydrophilic lining. Water molecules are strongly attracted and **adhere** to the hydrophilic walls, causing water to move up the vessel by capillarity. Adhesion also counters the effect of gravity on the water column.

> **Exam tip**
>
> Cohesion–tension theory is the only theory that can explain how water can reach the leaves of tall plants such as trees. It is therefore important that you know this theory in detail, as this will allow you to pick up the majority of the marks available on questions about the transport of water up a stem.

# Factors affecting the rate of transpiration

## Temperature

An increase in temperature will increase the kinetic energy of water molecules, which will **increase** the **rate of evaporation** from mesophyll cells, therefore increasing the rate of transpiration.

## Humidity

An increase in humidity will decrease the rate of transpiration as it will reduce the water potential **gradient** between the sub-stomatal air spaces and the external atmosphere. Figure 28 shows the difference in water potential between the tissues of the leaf, the sub-stomatal air spaces and the atmosphere outside the leaf.

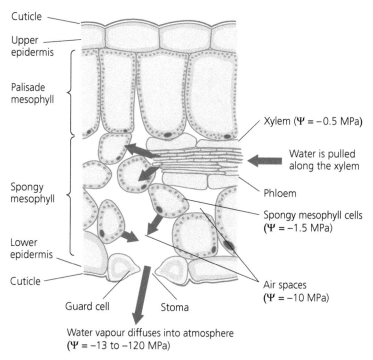

**Figure 28** How water moves through leaves

## Wind speed

An increased wind speed will increase the rate of transpiration. The wind disperses the water vapour around the stomata, which reduces the water potential outside the leaf so raising the water potential gradient between the sub-stomatal air spaces and the external atmosphere.

## Light intensity

An increase in light intensity will increase the rate of transpiration, as this will cause an opening of the stomata to allow $CO_2$ to enter the leaf for photosynthesis.

**Knowledge check 16**

Describe two environmental factors that can reduce the rate of transpiration in a plant.

**Exam tip**

The diameter of a tree trunk may be smaller during the middle of the day (when both the light intensity and temperature are high) than in the morning and in the evening. This is because the increased rate of transpiration produces negative pressure on the xylem vessels, which leads to a decrease in the vessels' diameter.

# Adaptations of plants to different habitats

## Xerophytes

Xerophytes are plants that have adapted to conditions of low water availability. They have adaptations to conserve water by **reducing transpiration**.

For example, the adaptations of marram grass (Figure 29) include:

- a thick, waterproof cuticle on the leaves that prevents evaporation through the cuticle
- stomata sunken into pits and hairs surrounding stomata that trap water vapour, which holds humid air around the stomata and so reduces the water potential gradient
- hinge cells, which cause the leaves to roll up and so retain humid air around the stomata, reducing the water potential gradient

**Exam tip**

Xerophytes are adapted to reducing the rate of transpiration mainly by reducing the water potential gradient between the sub-stomatal air spaces and the external atmosphere.

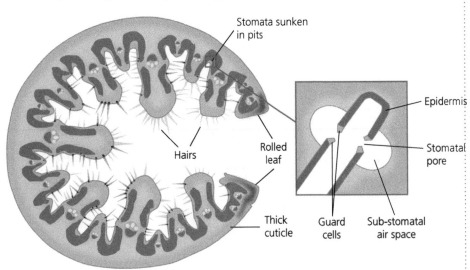

**Figure 29** Transverse section of marram grass

Other examples of xeromorphic adaptations in plants include:

- a rounded shape of the plant or leaf to reduce the surface area-to-volume ratio, reducing the rate of water loss
- the reduction of leaves to spines or needles, reducing the surface area for water loss
- the ability to fix $CO_2$ at night, so stomata can be closed during the day
- a shallow, extensive root system to absorb rainwater
- long taproots that can obtain water from sources deep underground

## Hydrophytes

Hydrophytes are plants that are adapted to **aquatic environments**. Water lilies, for example, live with their roots submerged in mud and have their leaves floating on the water surface. They are adapted in the following ways:

- stomata on upper epidermis of leaves to allow gas exchange
- a thin cuticle, as water loss is not problematic
- large air spaces in stem and leaf tissue to provide buoyancy
- poorly developed xylem, as the water provides support

**Knowledge check 17**

Hydrophytes, such as the water lily, do not have a waxy cuticle. True or false?

*Mesophytes*

Mesophytes are terrestrial plants that are adapted to neither a dry nor wet environment. Mesophytes grow well with an adequate water supply, while in prolonged dry periods they survive by shedding their leaves (to reduce transpiration), producing dormant seeds and surviving underground.

# The transport of organic compounds

Translocation involves the transport of organic molecules, mainly sucrose, from where they are produced (**sources**) to where they are utilised or stored (**sinks**).

## Evidence for translocation

Experimental evidence has shown that:

- sucrose is transported in the phloem
- sucrose is transported bi-directionally through the stem
- the rate of translocation is faster than diffusion

## Ringing experiments

Cylinders of bark can be removed from the plant stem — removing the phloem but leaving the xylem intact. After several days the composition of phloem above and below the ringed region was sampled. Sucrose and other organic solutes were present above the ring but absent below the ring. This indicates that translocation occurs in the phloem rather than in the xylem.

## Use of aphids and autoradiography

Aphids, which naturally feed on sap in the phloem, can be used to study the sap contents. Colonies of aphids can be allowed to feed on a plant. While they are feeding, they are anesthetised and their heads are cut off, leaving their mouthparts inserted into the phloem. The contents of the emerging sap, called the exudate, can then be studied.

If the plant is supplied with radioactive carbon dioxide ($^{14}CO_2$) and allowed to photosynthesise, then any carbohydrates it makes, including sucrose, will be radioactively labelled. The movement of these radioactive molecules in the shoot can be traced either by using photographic film or with a Geiger counter.

An **autoradiograph** is an image produced on photographic film — the presence of radioactive isotope will cause the film to 'fog', revealing the location of the labelled carbohydrates. This corresponds to the location of the phloem tissue in the stem.

## Mechanism for translocation

The exact mechanism for how organic molecules are transported in the phloem is not clear, but experiments have shown that it is too fast to be by diffusion alone. Different hypotheses have been put forward, including **mass flow** and **cytoplasmic streaming** — but neither of these can account for all the observations made.

**Knowledge check 18**

Explain why it is important that the xylem remains intact when setting up ringing experiments.

**Exam tip**

Although autoradiography involves the radioactive isotope of carbon ($^{14}C$), carbon itself is not transported through the plant. It is incorporated into organic molecules such as sucrose ($^{14}C_{12}H_{22}O_{11}$).

**Knowledge check 19**

Both water and sucrose are transported through a plant. Name the transport tissue for each, and then describe the direction in which sucrose and water travel through the stem.

*Mass flow hypothesis*

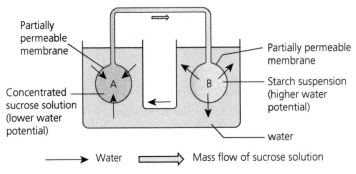

**Figure 30** Ernst Munch's mass flow model

The mass flow hypothesis suggests that translocation of solutes through the phloem occurs due to differences in hydrostatic pressure. The process of mass flow was demonstrated in an experiment that was set up by Ernst Munch in 1930 (Figure 30). The apparatus represents two cells, A and B, connected via a vessel. The two cells are surrounded by partially permeable membranes and immersed in water. Cell A represents a source cell, and contains a concentrated sucrose solution, and cell B represents a sink cell, and contains a starch suspension. Water enters cell A down a water potential gradient generated by the sucrose solution. This increases the hydrostatic pressure in cell A and forces the sucrose solution along the vessel to cell B, which has a lower hydrostatic pressure.

The mass flow theory suggests that the movement of solutes is **passive**. However, there are two sources of evidence that do not support this theory:

- the presence of large numbers of mitochondria in the companion cells
- translocation has only been observed in living phloem tissue

Both of these observations imply that translocation involves the use of ATP and is at least partly an active process.

*Pressure flow hypothesis*

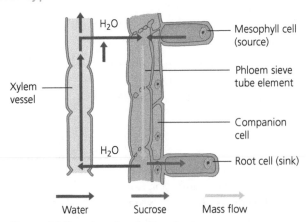

**Figure 31** Translocation in the plant

The pressure flow hypothesis suggests that translocation involves a combination of active transport and mass flow, and takes place in three stages:

1  *The loading of sucrose at the source*

Photosynthesis takes place in the leaf and produces glucose, which is converted to sucrose. The sucrose is **actively transported** from the leaf cells into the phloem sieve tubes. The loading is carried out by companion cells, which contain many mitochondria.

2  *The translocation of sap from source to sink*

The presence of sucrose in the sieve tube lowers the water potential, causing water to enter the phloem, from the xylem, by osmosis. As the hydrostatic pressure in the phloem sieve tube increases, **mass flow** occurs and the sap moves through the phloem. Pores in the sieve plates allow large molecules to pass through, which enables unrestricted mass flow.

3  *The unloading of sucrose at the sinks*

At the sinks the sucrose is actively transported out of the phloem and is either converted to starch and stored or converted to glucose and respired. The loss of sucrose from the phloem increases the water potential causing water to pass out of the phloem, reducing the hydrostatic pressure and **maintaining the pressure gradient** from source to sink.

One criticism of the pressure flow theory is that substances cannot flow in opposite directions in the same sieve tube. However, phloem contains numerous sieve tubes and different solutes could travel in opposite directions at the same time by travelling in two different sieve tubes. Translocation in different sieve tubes could also account for different solutes travelling at different rates.

Another hypothesis, known as **cytoplasmic streaming**, may also be involved in translocation. Within individual sieve tube elements, the cytoplasm circulates around the cell. Solutes may then be actively transported across the sieve plate, either upwards or downwards in the sieve tube. If this hypothesis is correct it would account for the movement of solutes in opposite directions in the same sieve tube.

## Practical work

### Measuring the rate of transpiration

The rate of transpiration can be investigated using a **potometer** (see Figure 32).

An air bubble is introduced into the open end of the apparatus and the movement of the bubble in a given time period can be recorded. This will measure the rate of water uptake by the plant, which will give a close approximation of the rate of transpiration.

Different environmental factors can affect the rate of transpiration, and this provides examiners with opportunities to assess a range of practical and mathematical skills. These could include:

■ precautions in setting up the potometer
■ the identification of controlled variables
■ drawing graphs and calculations of volumes and rates
■ refining the investigation to obtain more accurate or additional data

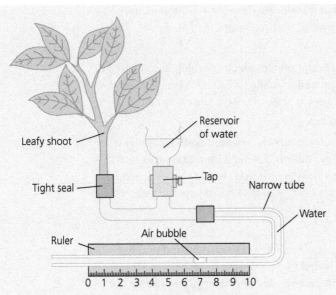

**Figure 32** A typical school potometer for measuring the rate of water uptake by leafy shoots

Cohesion-tension theory relies on there being an unbroken column of water in the xylem vessel; therefore it is important that no air enters the xylem of the shoot or the apparatus. When setting up the investigation the leafy shoot is cut underwater and placed into the bung of the potometer, which is filled with water. Vaseline can then be used to ensure that the seal between the bung and the shoot is airtight.

Transpiration rate is affected by stomatal density as well as environmental factors such as temperature, humidity, wind speed and light intensity. If one of these factors is the independent variable in an investigation, it is important that all of the other factors are controlled.

You may be provided with data on the distance travelled by the air bubble in a fixed period of time, as well as the internal diameter of the narrow tube. From this information you could be asked to calculate the rate of water uptake. For example, if the bubble moves 77 mm in 2½ minutes and the internal diameter of the tube is 0.8 mm, then the rate of uptake can be calculated as follows:

- if the internal diameter is 0.8 mm then the radius is 0.4 mm
- the area (of a circle) = $\pi r^2$, so the cross-sectional area of the water column = $3.14 \times 0.4^2 = 0.502\,\text{mm}^2$
- the volume of water uptake = $0.502 \times 77 = 38.68\,\text{mm}^3$
- rate of water uptake = $38.68/150$ seconds = $1.45\,\text{mm}^3\,\text{s}^{-1}$

You may be provided with a method that a student used to obtain their data. You could be asked to identify any limitations in the method that may have led to inaccurate results, and to suggest how the method could be improved. These limitations might be due to the precautions not being followed, or to the accuracy of the scale used. You could also be asked to adapt the method →

**Knowledge check 20**

Using the data on the left, state the rate of transpiration in:

a $\text{cm}^3\,\text{s}^{-1}$
b $\text{cm}^3\,\text{h}^{-1}$

Give your answers to three significant figures.

to investigate a different variable, that is, to change the independent variable — it is important to ensure that the original independent variable would now become a controlled variable.

During this topic you will also have examined microscope slides and electron micrographs of tissues from the leaves, roots and stems of a dicotyledonous plant. In the exam you may be given photomicrographs or drawings of these tissues. It is important that you can identify the various structures and link them to their functions.

You will also have examined the leaves of marram grass (a xerophyte) and water lily (a hydrophyte) to help you to better understand the adaptations of plants to particular environments. In the exam you may be given photomicrographs of these or other species and be asked to identify how these species are adapted to the environments in which they are found.

## Summary

After studying this unit you should be able to demonstrate and apply your knowledge and understanding of:
- the structure of the root and primary stem from a dicotyledon, including the role of the endodermis
- the absorption of water by the root and the movement of water through the apoplast, symplast and vacuolar pathways
- the detailed structure of xylem and phloem as seen under both light and electron microscopes
- the different theories to explain the movement of water from root to leaf, including cohesion-tension theory
- the effect of environmental factors that affect transpiration and the adaptations shown by some angiosperms — xerophytes and hydrophytes — which are adapted to different environmental conditions
- the evidence to support the translocation of organic materials from source to sink, including the use of aphids and autoradiographs, and that explanations of the mechanism for translocation are contentious

# Transport in animals

As organisms get larger their surface area-to-volume ratio decreases and diffusion distances increase. Large organisms require vascular (transport) systems because diffusion is inefficient over large distances. **Mass transport** is the bulk movement of substances through a transport system using force.

## Types of vascular system

There are two basic types of vascular system in animals:
- In an **open circulation** the blood is not confined to vessels. For example, in insects the blood is pumped by a dorsal, tube-shaped heart, but it flows freely over the tissues, through spaces that are collectively known as haemocoel. The blood flows slowly, at low pressure, and there is little control over its distribution.

- In a **closed circulation** the heart pumps blood, under high pressure, through **blood vessels**; organs are not in direct contact with the blood. Respiratory gases are transported in blood by a respiratory pigment, such as **haemoglobin**. Annelids and vertebrates have a closed circulation.

Figure 33(a) shows a **single circulation** in a fish:
- Blood passes through the heart once every time it passes around the body.
- Blood leaves the heart under high pressure, but the pressure must fall before it reaches the gill capillaries.
- Blood then flows slowly, under low pressure, around the rest of the body before returning to the heart.

Figure 33(b) shows a **double circulation** in a mammal:
- Blood must pass through the heart twice every time it goes round the body once.
  - The **pulmonary circulation** transports blood between the heart and the lungs.
  - The **systemic circulation** transports blood between the heart and all the other organs of the body.
- The blood flows quickly, under high pressure, in both the pulmonary and systemic circulations.

> **Knowledge check 21**
>
> An insect's blood lacks haemoglobin. Explain why this is not a disadvantage to gas exchange in the insect.

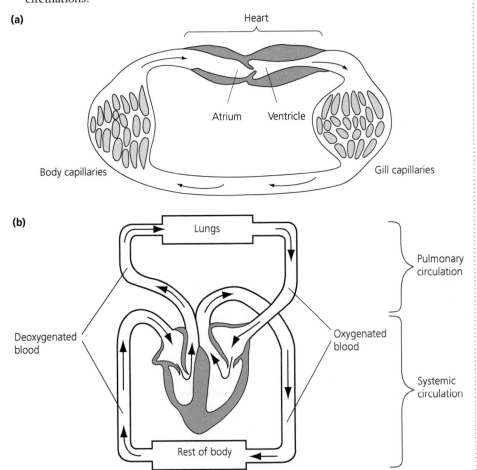

**Figure 33** (a) The fish circulatory system; (b) the mammalian circulatory system

# The mammalian heart

The mammalian heart is a four-chambered **pump** that is situated in the thoracic cavity (Figure 34). It is a specialised organ and consists mainly of **cardiac muscle**. The structure of the heart allows the complete separation of oxygenated and deoxygenated blood.

Table 9 shows the functions of the different structures in the mammalian heart.

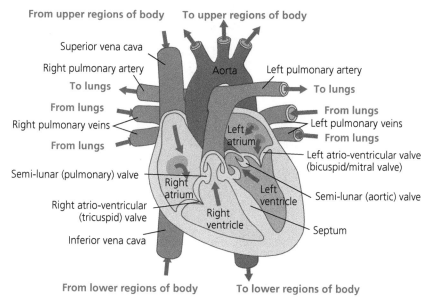

**Figure 34** The internal structure of the heart

| Structure | Function |
|---|---|
| Pulmonary artery | Carries deoxygenated blood to lungs from right ventricle |
| Pulmonary vein | Carries oxygenated blood from lungs to left atrium |
| Aorta | Carries oxygenated blood from left ventricle to the body tissues |
| Vena cava | Carries deoxygenated blood from body to right atrium |
| Atria | Thin-walled chambers, which receive blood |
| Ventricles | Chambers with thick walls that generate high pressure of blood when the walls contract, so that the blood can be forced over a great distance. The left ventricle is larger than the right ventricle, and has a thicker muscular wall that can generate a higher pressure so that blood can travel the greater distance to the extremities of the body |
| Atrio-ventricular valves | Prevent backflow of blood from the ventricles to the atria during ventricular systole |
| Valve tendons ('heart strings') | Keep valves under tension and prevent them from inverting during ventricular systole |
| Semi-lunar valves | Prevent backflow of blood from the arteries to the ventricles (the only examples of valves in arteries) |

**Table 9** The functions of the different structures in the mammalian heart

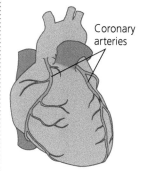

**Figure 35** The position of the coronary arteries

# The cardiac cycle

Figure 36 shows the sequence of events that take place during the cardiac cycle.

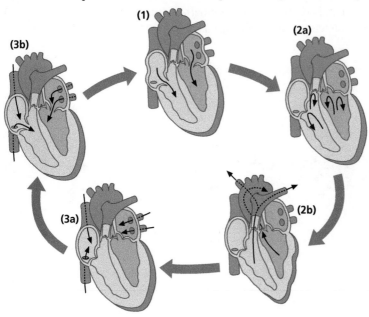

**Figure 36** The stages of the cardiac cycle

*Atrial systole*

■ As the atria contract, atrial pressure exceeds ventricular pressure, causing the atrio-ventricular valves to open and blood flows from the atria into the ventricles (stage 1).

*Ventricular systole*

■ As the ventricles contract, the ventricular pressure exceeds atrial pressure, causing the atrio-ventricular valves to close, which generates the first heart sound — 'lub' (stage 2a).
■ When ventricular pressure exceeds aortic pressure, the semi-lunar valves open and blood flows from the ventricle into the arteries (stage 2b).

*Atrial and ventricular diastole*

■ As the ventricles relax, ventricular pressure falls below pressure in the arteries, causing the semi-lunar valves to close, which generates the second heart sound — 'dup'.
■ Low-pressure blood in the veins returns to the heart as the atria relax (stage 3a).
■ When ventricular pressure falls below atrial pressure, the atrio-ventricular valves open again and blood flows from the atria into the ventricles (stage 3b).

**Exam tip**

When describing any of the phases in the cardiac cycle you must make reference to:

■ which chamber is contracting and which is relaxing, and relate this to the relative pressures in both chambers
■ which valve is open and which valve is closed

The cardiac cycle can be analysed graphically, as shown in Figure 37.

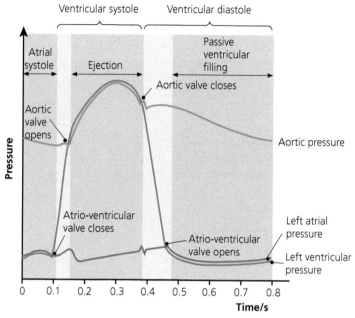

**Figure 37** Pressure changes associated with the cardiac cycle

**Knowledge check 24**

Explain the difference between the minimum blood pressure in the left ventricle and in the aorta.

# The cardiac impulse

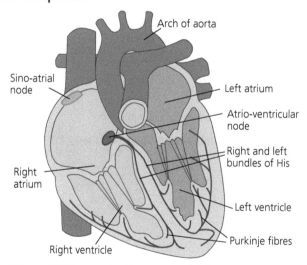

**Figure 38** The conducting system of the heart includes the SAN, AVN and Purkinje fibres

- Cardiac muscle is myogenic.
- The cardiac impulse (heart beat) originates in the **sino-atrial node** (**SAN**), which acts as a **pacemaker** and is located in the wall of the right atrium (Figure 38).
- The impulse spreads out into the walls of the atria, causing **atrial systole**.
- There is a layer of non-conductive connective tissue that prevents the impulse from travelling down the wall of the ventricles.

Myogenic muscle can initiate its own contraction, that is, it does not need to be stimulated by a nerve to cause it to contract.

- The impulse is picked up by the **atrio-ventricular node** (**AVN**), which is also located in the wall of the right atrium, at its base.
- The impulse is then conducted through the **bundle of His** to the apex of the ventricles.
- The impulse then travels upwards through the branching **Purkinje fibres**, through the walls of the ventricles, causing the cardiac muscle of the ventricles to contract (**systole**) *from the bottom upwards.*

### Electrocardiogram (ECG)

The electrical activity that spreads through the heart during the cardiac cycle can be detected using electrodes placed on the skin. The electrical signals can then be shown on a cathode ray oscilloscope or a chart recorder. The record produced by this procedure is called an electrocardiogram (ECG).

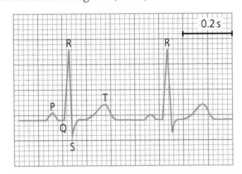

**Figure 39** An electrocardiogram

A typical ECG consists of characteristic patterns, or waves, which correspond to particular events in the cardiac cycle (Figure 39).

- The P wave shows the depolarisation of the atria during atrial systole.
- The QRS wave shows the spread of depolarisation through the ventricles during ventricular systole.
- The T wave shows the repolarisation of the ventricles during ventricular diastole.

ECGs can be used to calculate heart rate, by measuring the time interval between the same points of successive cycles on the trace.

Doctors can also use changes in the pattern of an ECG to help diagnose cardiovascular disease and heart defects.

### Control of cardiac activity

At rest the heart beats at approximately 75 beats per minute. During periods of activity/inactivity the heart rate can be varied by nerves that originate from the cardiovascular centre in the medulla of the brain. Impulses can be sent via different neurons to either stimulate the SAN, which increases the heart rate, or to inhibit the SAN and decrease the heart rate. These changes to heart rate are **involuntary** and are not under conscious control.

The hormone **adrenalin**, which is released when your body is under stress, also causes the SAN to discharge at a higher frequency, increasing heart rate.

**Knowledge check 25**

There is a short time delay between the impulse arriving at the AVN and leaving the AVN. What is the advantage of this?

**Knowledge check 26**

Use the ECG in Figure 39 to calculate the heart rate.

# Blood vessels

There are five types of blood vessel (Figure 40):

- **Arteries** transport blood away from the heart.
- **Arterioles** connect arteries to capillaries.
- **Capillaries** are microscopic vessels that form networks within the tissues of the body.
- **Venules** connect capillaries to veins.
- **Veins** transport blood back to the heart.

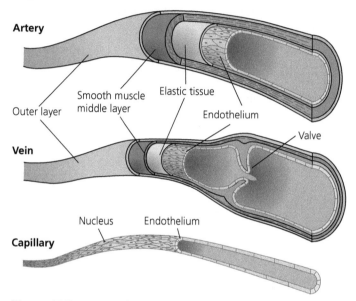

**Figure 40** Structure of an artery, a vein and a capillary

| Feature | Artery | Vein | Capillary |
|---|---|---|---|
| | 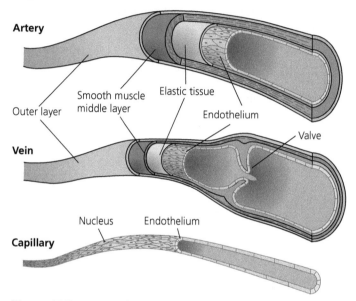 | | |
| Outer (protective) layer | Present with collagen fibres | Present with collagen fibres | Absent |
| Middle layer | Thick layer with smooth muscle and many elastic fibres | Thin layer with smooth muscle and few elastic fibres | Absent — no muscle or elastic tissue |
| Endothelium (smooth to reduce friction) | Present | Present | Present |
| Pressure | High | Low | Low |

**Table 10** Summary of the main structural differences between arteries, veins and capillaries

## Arteries and arterioles

As a result of ventricular systole blood enters the large arteries under high pressure; arteries have thick walls to resist this pressure. They contain many elastic fibres, which stretch to allow the large arteries to bulge to accommodate the blood.

The pressure in these arteries shows a rhythmic rise and fall (which corresponds to ventricular systole and diastole), generating a pulsatile flow (Figure 41). **Elastic recoil** of the elastic fibres pushes blood along the artery and produces an even flow of blood through the arteries.

In the smaller arteries and arterioles the smooth muscle fibres can contract and relax, which can change the diameter of the lumen and therefore help to control blood-flow to different parts of the body:

- When the smooth muscle contracts this causes vasoconstriction (the lumen becomes narrower) and restricts the blood-flow.
- When the smooth muscle relaxes this causes vasodilation (the lumen becomes wider) and increases the blood-flow.

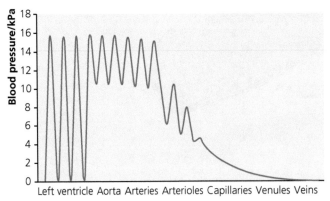

**Figure 41** The changes in blood pressure at different places in the systemic circulation during ventricular systole and diastole (not drawn to scale)

As the blood travels further from the heart the blood vessels become increasingly branched, and the blood pressure and the rate of blood flow decrease. This is due to:

- the increase in total cross-sectional area
- frictional resistance of blood flowing along the blood vessels

## Capillaries

The **millions** of capillaries in our bodies act as **exchange surfaces** (Figure 42). Table 11 shows the characteristics of capillaries and how they are adapted to carry out exchange of materials.

**Knowledge check 27**

Explain the importance of vasoconstriction and vasodilation to a mammal.

**Exam tip**

The aorta has a high proportion of elastic fibres; during ventricular systole they allow the aorta to stretch to accommodate the blood entering, then during diastole they recoil to maintain blood pressure and force blood along the artery. Do not confuse the properties of elastic fibres with those of muscle — elastic fibres cannot contract, they can only stretch and recoil.

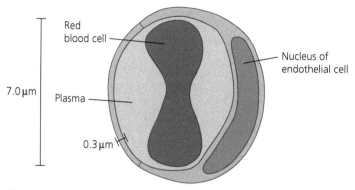

**Figure 42** Cross-section of a capillary

| Structural adaptation | Advantage |
|---|---|
| Form a dense network | Large surface area for diffusion |
| Capillary walls are one cell thick and flattened | Short diffusion pathway |
| Small diameter/narrow | |
| Red blood cells in contact with wall | |
| Narrow lumen | Reduces flow rate as red blood cells pass singly, giving more time for diffusion |
| Capillary walls are permeable to gases | $O_2$ and $CO_2$ can easily diffuse into/out of the blood |
| Capillary walls contain pores between the endothelial cells | Allows the formation of tissue fluid |

**Table 11** The characteristics of a capillary

## Venules and veins

The return of blood to the heart is non-rhythmic. **Semi-lunar valves** in the veins prevent backflow. Although the pressure in veins is low, blood is returned to the heart due to:

- pressure increases caused by skeletal muscle (see Figure 43):
  - When skeletal muscles (surrounding the vein) contract they squeeze the vein, reducing the volume and increasing the pressure inside the vein.
  - This forces blood through the valve 'in front of' the blood and causes the valve behind to close, so preventing backflow.
  - This ensures that blood travels in one direction only.
- residual pressure of blood during ventricular systole
- negative pressure due to atrial diastole (suction effect)
- negative pressure in the thorax during inspiration (suction effect)

**Knowledge check 28**

Figure 43 shows a vein surrounded by skeletal muscle and bone. When the muscle contracts state the direction in which blood will flow, which valve will be open and which valve will be closed.

**Exam tip**

When answering questions on blood vessels it is always important to link the structure of the vessel with its function.

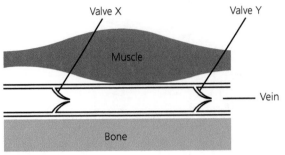

**Figure 43** A vein surrounded by bone and skeletal muscle

# Blood

Blood is made up of **plasma** (55%) and **cells** (45%). Plasma consists of 90% water; the other 10% is made up of solutes, for example glucose and amino acids. Plasma proteins are also dissolved in plasma and are important in the formation of tissue fluid. The function of plasma is to transport the soluble products of digestion, as well as ions, hormones, antibodies and excretory products such as urea. It also distributes heat around the body.

**Red blood cells** are responsible for transporting respiratory gases around the body.

**White blood cells** are involved in protecting the body against infection — immunity.

**Platelets** are involved in blood clotting.

# The formation of tissue fluid

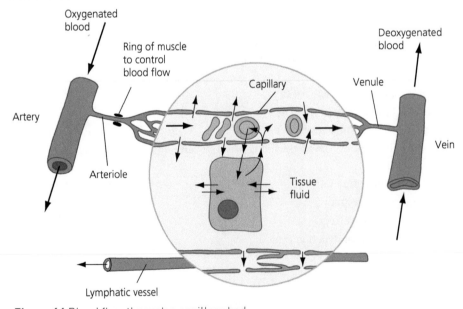

**Figure 44** Blood flow through a capillary bed

The capillary wall is made up of a single layer of endothelial cells. There are minute pores between the individual cells that allow solutes to be exchanged between the blood and cells of the tissues.

At the arterial end of the capillary the hydrostatic pressure is relatively high (caused by ventricular systole) and this causes ultrafiltration to occur:

- Fluid (water and small solutes) is forced out of the blood.
- The larger plasma proteins and cells remain in the blood as they are too big to pass through the pores (Figure 44).

The fluid forced out of the blood is now called **tissue fluid** and it bathes the cells in essential nutrients. As fluid is forced out of the capillary the hydrostatic pressure in the capillary falls.

At the venule end of the capillary the water potential of the blood is lower than that of the tissue fluid due to the soluble plasma proteins. This causes water to be drawn back into the capillary by osmosis.

Figure 45 shows that at the arteriole end of the capillary the hydrostatic pressure is greater than the solute potential of the plasma, so fluid is forced out into the tissues. At the venule end of the capillary the solute potential of the plasma is greater than the hydrostatic pressure, so water is drawn back into the blood by osmosis.

> Solute potential refers to the effects of solutes lowering the water potential of a system, which acts to draw water into a system.

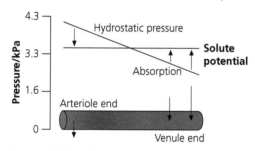

**Figure 45** Relationship between hydrostatic pressure and solute potential

More fluid leaves the capillaries than is reabsorbed. Excess tissue fluid drains into blind-ended capillaries of the **lymphatic system**. The lymph capillaries join to form lymph vessels, which return the lymph to the blood via the thoracic duct. Lymph vessels also have valves to ensure one-way flow.

## Exam tip

It is easy to lose marks in the exam for incorrect terminology. Many students make reference to blood being forced out of the capillary; this is incorrect, *fluid* is forced out at the arteriole end of the capillary. You should refer to *water* moving back into the capillary at the venule end, because osmosis is involved.

## Knowledge check 29

Describe how fluid that has been forced out of the capillary is returned to the blood.

# Transport of oxygen

Oxygen is transported around the body in red blood cells. Red blood cells have many adaptations to increase the efficiency of this transport. They have a large surface area-to-volume ratio because:

- they are very small (about $8\,\mu m$ in diameter)
- they lack a nucleus, giving them a biconcave disc shape

The lack of a nucleus and other organelles means that more haemoglobin (Hb) can be contained within the cytoplasm of the cell. Haemoglobin is a globular protein that has a quaternary structure (Figure 46). It is made up of four polypeptide chains — each contains a haem group. The haem group contains iron ions ($Fe^{2+}$), which form a loose association with $O_2$ molecules. As there are four haem groups, each molecule of haemoglobin can transport four molecules of oxygen.

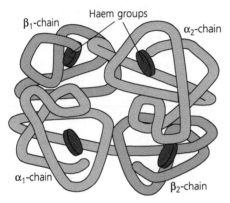

**Figure 46** The quaternary structure of haemoglobin

When haemoglobin combines with $O_2$ it is called oxyhaemoglobin — this reaction is reversible:

$$Hb \quad + \quad 4O_2 \quad \rightleftharpoons \quad HbO_8$$

Haemoglobin                 Oxyhaemoglobin

Haemoglobin should be thought of as an 'O$_2$ taxi'. It picks up, that is, **loads** oxygen in the lungs and drops off, that is, **unloads** $O_2$ at the tissues where it is needed for respiration.

## The oxygen–haemoglobin dissociation curve

The oxygen–haemoglobin dissociation curve (Figure 47) shows the relationship between the partial pressure of oxygen ($pO_2$) and the percentage saturation of haemoglobin with $O_2$.

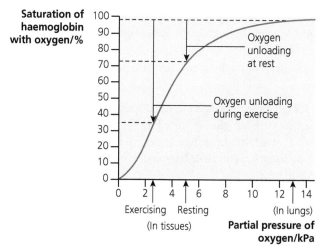

**Figure 47** Oxygen dissociation curve of human haemoglobin

The shape of the curve is significant.

- The curve is S-shaped, which indicates that haemoglobin is efficient at loading $O_2$ and can become fully saturated at a lower $pO_2$ than if the relationship was a linear one.
- Haemoglobin has a high affinity for $O_2$ at a relatively high $pO_2$ and will therefore load $O_2$ to form oxyhaemoglobin. This occurs in the capillaries of the lungs.
- Haemoglobin has a low affinity for $O_2$ at a low $pO_2$ and will therefore unload oxygen. This occurs in the capillaries of the tissues.
- The steep part of the curve means that for a small decrease in the $pO_2$ there will be a large decrease in the percentage saturation of haemoglobin with $O_2$. This means that more $O_2$ will be unloaded to the tissues to be used in aerobic respiration.

From Figure 47 you can see that for a drop in $pO_2$ of 2.5 kPa (tissues at rest to tissues exercising) the percentage saturation of haemoglobin falls by about 40%, that is, 40% more oxygen is unloaded to the tissues for respiration.

*Fetal haemoglobin and other respiratory pigments*

Figure 48 shows that the oxygen dissociation curve for fetal haemoglobin is situated to the left of that for 'normal' adult haemoglobin.

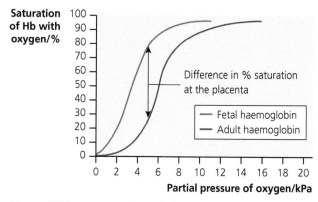

**Figure 48** The oxygen dissociation curves of fetal and adult haemoglobin

**Exam tip**

Many students tend to focus on haemoglobin's ability to load $O_2$. However, it is the unloading of $O_2$ by haemoglobin that is crucial — $O_2$ is unloaded to the tissues so that they can carry out aerobic respiration.

Fetal haemoglobin has a higher affinity for oxygen — at any given $pO_2$ the percentage saturation of haemoglobin is higher. The significance of this is that the maternal haemoglobin will unload approximately 70% of its $O_2$ to the tissues of the placenta; the fetal haemoglobin will then load $O_2$ from the placenta to become approximately 80% saturated.

Organisms that live in low $O_2$ environments, for example, llamas (at high altitude) and lugworms (in burrows in sand), also have pigments with a higher affinity for $O_2$ — that is, with a dissociation curve that is to the left of adult human haemoglobin. The advantage of this is that these pigments can load more $O_2$ and their haemoglobin can become fully saturated at a lower $pO_2$.

**Knowledge check 30**

Figure 49 shows the dissociation curve for myoglobin, the respiratory pigment found in muscle fibres. Use the diagram to explain the function of myoglobin.

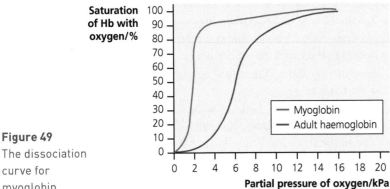

**Figure 49**
The dissociation
curve for
myoglobin

## The Bohr effect

During exercise the muscles are working harder, so they need more ATP. Therefore the rate of respiration increases. This also produces more $CO_2$, which lowers the pH of the blood. This causes the dissociation curve to shift to the right — the Bohr effect (Figure 50).

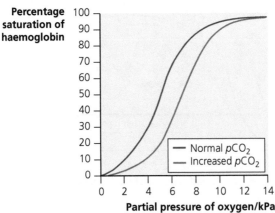

**Figure 50** The Bohr effect

The haemoglobin now has a lower affinity for $O_2$ — at any given $pO_2$ the percentage saturation of haemoglobin is lower. The advantage is that haemoglobin unloads more $O_2$ to the muscles tissues for an increased rate of respiration.

# Transport of carbon dioxide

Carbon dioxide is transported in the blood in three different ways. About 5% dissolves directly into the plasma, and a small percentage combines with amino groups in haemoglobin to form carbaminohaemoglobin. The majority of $CO_2$ is transported as **hydrogencarbonate ions** ($HCO_3^-$) in the plasma.

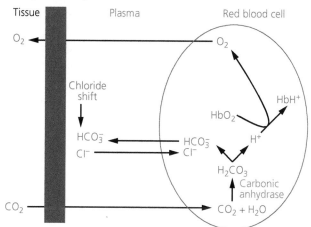

**Figure 51** A red blood cell showing the biochemical pathway involved in transporting $CO_2$

$CO_2$ diffuses into the red blood cell and combines with $H_2O$ to form carbonic acid ($H_2CO_3$). This reaction is catalysed by the enzyme **carbonic anhydrase** (Figure 51). The carbonic acid then dissociates into protons ($H^+$) and hydrogencarbonate ions ($HCO_3^-$). The $H^+$ will lower the pH of the cell and so it needs to be buffered — this is done by haemoglobin. The oxyhaemoglobin dissociates, unloading oxygen, which diffuses out of the red blood cell and into the tissues. $HCO_3^-$ that has accumulated in the red blood cell diffuses out into the plasma.

Electroneutrality is maintained by the inward diffusion of chloride ions from the plasma. This is known as the **chloride shift**.

In the capillaries of the lungs all of these reactions occur in reverse.

## Practical work

During this topic you will have made a dissection of a heart, which will have helped you to better understand how heart structure relates to the pumping of blood during the cardiac cycle. You will have also examined microscope slides of erythrocytes and transverse sections through an artery and a vein. In the exam you may be given photomicrographs or drawings of these sections. It is important that you can identify the various tissues and link them to their functions.

## Summary

After studying this topic you should be able to demonstrate and apply your knowledge and understanding of:

- the similarities and differences of the vascular systems of insects, earthworms, mammals and fish
- the structure and function of the heart and blood vessels of the mammalian circulatory system, including the names of the main blood vessels associated with the human heart
- the pressure changes that occur during the cardiac cycle and the role of the sino-atrial node and Purkinje fibres in controlling the cardiac cycle
- the structure and function of blood with regard to:
  - the transport of nutrients, hormones, excretory products and heat
  - the formation of tissue fluid
  - the transport of respiratory gases
- oxygen–haemoglobin dissociation curves, including a comparison of the dissociation curves in animals that are adapted to different environmental conditions
- the effects of exercise on the oxygen–haemoglobin dissociation curve

# Adaptations for nutrition

Organisms are referred to as either autotrophs or heterotrophs, depending on their nutrition.

Plants are autotrophs. They convert carbon dioxide ($CO_2$) and water ($H_2O$) into glucose ($C_6H_{12}O_6$) using light energy during photosynthesis. Plants are referred to as photoautotrophs as they use light as their source of energy. Some species of bacteria are also autotrophic; however, they obtain their energy from exergonic chemical reactions. They are therefore referred to as chemoautotrophs.

Heterotrophs obtain nutrition from complex organic molecules. These complex organic molecules tend to be large, insoluble molecules, such as polysaccharides (like starch) and proteins, which cannot cross plasma membranes. Chemical digestion using enzymes is necessary to hydrolyse these molecules into small, soluble molecules, such as glucose and amino acids. These small, soluble molecules can then be **absorbed** and **assimilated** into other molecules that the organism requires. Many animals, including humans, consume complex organic molecules and break them down in the gut. This is known as **holozoic** nutrition. **Saprophytes** carry out extracellular digestion to obtain nutrition from dead organic matter, that is, from dead organisms and animal waste (urine and faeces).

## Heterotrophic nutrition

### Unicellular organisms

Unicellular organisms, for example amoeba (kingdom Protoctista), ingest food particles, which may include bacteria and algae, by phagocytosis and carry out **intracellular** digestion.

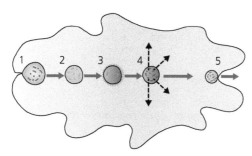

**Figure 52** Intracellular digestion in an amoeba

Amoebas are mobile organisms that can engulf other unicellular organisms, such as bacteria (Figure 52). Ingestion occurs via phagocytosis, a form of endocytosis by which the bacteria are engulfed by the plasma membrane of the amoeba (stage 1 in Figure 52) and enclosed within a vacuole (stage 2). Lysosomes then migrate and fuse with the vacuole membrane, releasing hydrolytic enzymes into the vacuole (stage 3). Intracellular digestion takes place and the large, insoluble molecules in the bacteria are broken down into small, soluble molecules. The soluble products of digestion are then passed out of the vacuole into the cytoplasm of the amoeba by diffusion and active transport (stage 4). The undigested material is carried by the vacuole to the cell membrane and removed from the amoeba by exocytosis (stage 5).

**Autotrophs** are organisms that synthesise complex organic molecules from simple inorganic molecules, using a source of energy.

**Heterotrophs** are organisms that obtain nutrition from complex organic molecules.

**Exam tip**

Synoptic link to Unit 1: the example of digestion in amoeba demonstrates the processes of endocytosis and exocytosis (the movement of large insoluble molecules across membranes) and could be used to answer questions in both the Unit 1 and Unit 2 papers.

# Multicellular organisms

*Hydra* is a multicellular animal of the phylum Cnidaria, which includes sea anemones, jellyfish and corals. *Hydra* feeds primarily on water fleas (*Daphnia*) and has a simple, undifferentiated, sac-like gut with a single opening, which acts as both a mouth and an anus (Figure 53).

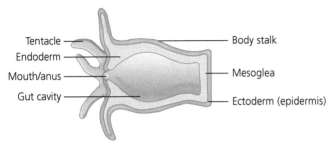

**Figure 53** Longitudinal section of *Hydra* showing an undifferentiated, sac-like gut

The tentacles of *Hydra* trap water fleas that brush against them. The tentacles then bend over towards the single opening, which acts as a 'mouth' and opens wide enough to allow the prey to enter the single gut cavity (called the enteron). The endodermis secretes hydrolytic enzymes into the gut, which initiate **extracellular** digestion. The extracellular enzymes do not break down the food completely. The partially digested food is transported into the endodermal cells via phagocytosis and is digested intracellularly, as in *Amoeba*. The undigested material is egested from the gut via the single opening.

> **Knowledge check 32**
>
> Describe the difference between intracellular digestion and extracellular digestion.

# Structure of the human alimentary canal (gut)

Mammals, including humans, have a much more complex digestive system than *Hydra*. Instead of having a sac-like gut with a single opening, they have a long tube, the **alimentary canal**, which runs through the body from the mouth to the anus.

In mammals, holozoic nutrition can be broken down into four stages:

1 **ingestion**
   - taking in food

2 **digestion**
   - **mechanical digestion** reduces the size of the food material, increasing its total surface area and making chemical digestion more efficient
   - **chemical digestion** involves the hydrolysis of large, insoluble molecules (polymers) into small, soluble molecules (monomers), using **enzymes**. Chemical digestion in humans is an example of **extracellular** digestion, as the enzymes involved are secreted into the lumen of the gut

3 **absorption**
   - the passage of small, soluble molecules (monomers) and other useful substances into the bloodstream. These molecules then pass into the body cells where they are either respired to release energy, or used to synthesise other molecules that the cell needs, for example enzymes. This is known as **assimilation**

4 **egestion**
   - the elimination of undigested food material from the body

> **Exam tip**
>
> Do not confuse egestion, which is the elimination of material from a body cavity, with excretion, which is the elimination of the waste products of metabolism from within the body's cells.

The human gut is adapted to a mixed, omnivorous diet that includes both plant and animal material. As a consequence of this the human gut is divided into several specialised regions for the efficient digestion of different food substances (Figure 54).

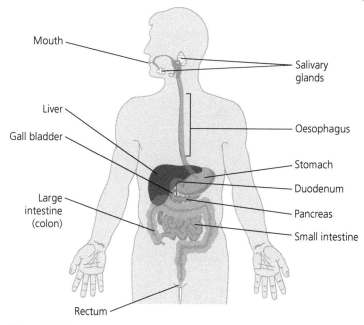

**Figure 54** The human gut

The gut wall has the same layers throughout, but they are adapted, in different regions, to carry out particular functions. The basic structure is shown in Figure 55.

Table 12 summarises the functions of the different tissues in the gut wall.

External organs such as the **salivary glands, pancreas** and **liver** also provide digestive secretions, which enter the lumen of the gut via ducts.

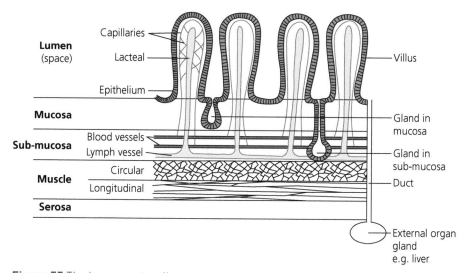

**Figure 55** The human gut wall

| Epithelium | Single layer of cells lining the gut wall |
|---|---|
| Mucosa | Contains glands that produce digestive secretions:<br>■ Digestive enzymes<br>■ Acidic/alkaline fluid to provide the optimum pH for digestive enzymes<br>■ Mucus (secreted by goblet cells) to provide lubrication and protection to the gut wall |
| Sub-mucosa | ■ Contains blood vessels and lymph vessels for the transport of the soluble products of digestion<br>■ Contains glands that produce alkaline fluid |
| Circular and longitudinal muscles | Antagonistic muscle pair — bring about peristalsis due to contraction of the circular muscle behind the bolus of food |
| Serosa | Tough, protective coat surrounding the gut |

**Table 12** The functions of the different tissues in the gut wall

**Exam tip**

You are not expected to know the names of the different glands, or the names of specialised cells within the wall of the gut.

# Regions of the gut

## The mouth/buccal cavity (pH 6.5–7.5)

- **Mechanical digestion** occurs — using the teeth to increase the surface area of food by reducing its size.
- **Chemical digestion** of **starch** occurs.
- Salivary glands produce **amylase, mineral ions** and **mucus**.

## The oesophagus

- Transfers the bolus of food from the buccal cavity to the stomach by **peristalsis**, which occurs as a wave of muscular contraction. During peristalsis the circular muscle of the oesophagus contracts behind the bolus of food, while the longitudinal muscles relax in front of it. This combined action forces the bolus towards the stomach.
- **Goblet cells** in the mucosa secrete **mucus** to provide lubrication.

## The stomach (pH 2)

- **Mechanical digestion** occurs. There are three layers of muscle that churn food into liquid chyme.
- Chemical digestion of **proteins** begins.
- Gastric glands secrete:
  - endopeptidases to hydrolyse peptide bonds in the proteins
  - HCl, which activates and provides the optimum pH for the endopeptidases, and to kill any bacteria that have been ingested
  - alkaline mucus to protect the stomach wall from the hydrochloric acid and endopeptidases

## The small intestine: the duodenum and ileum (pH 7–8)

- Main site of **chemical digestion**.
- **Intestinal glands** in the wall of the duodenum produce digestive **enzymes** and **alkaline fluid**. The **pancreas** also secretes enzymes and alkaline fluid into the duodenum via the pancreatic duct. The alkaline fluid neutralises the acidic chyme from the stomach and provides the optimum pH for the digestive enzymes.
- The ileum is the main site of **absorption** of the soluble products of digestion.

*The large intestine: colon and rectum*

- The colon absorbs water and vitamins.
- The rectum is for the temporary storage of faeces.

# Chemical digestion in the mammalian gut

During chemical digestion large polymer molecules are hydrolysed into smaller monomers, as shown in Figure 56.

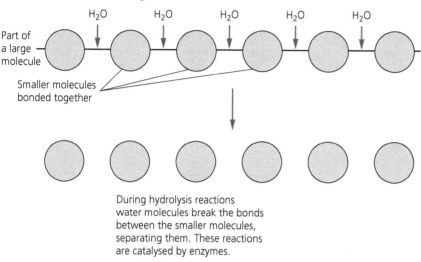

**Figure 56** Digestion involves the hydrolysis of polymers into monomers

In mammals, the breakdown of food occurs by both extracellular digestion and intracellular digestion. The salivary glands, gastric glands and pancreas all secrete enzymes into the lumen of the gut where extracellular digestion occurs. The enzymes that hydrolyse disaccharides into monosaccharides and dipeptides into amino acids are found in the outer membranes and the cytoplasm of epithelial cells lining the small intestine where intracellular digestion occurs.

## Chemical digestion of carbohydrates

- Starch is **hydrolysed** into maltose by **amylase** (optimum pH 8).
- Amylase is found in **saliva** and in **pancreatic juice** (which is released into the duodenum).
- Maltose is then hydrolysed to glucose by **maltase**.
- Sucrose is hydrolysed to glucose and fructose by **sucrase**.
- Lactose is hydrolysed into glucose and galactose by **lactase**.
- Maltase, sucrase and lactase are located on the **cell surface membrane** of the epithelial cells of the small intestine.

## Chemical digestion of lipids

- Lipids are broken down mechanically by **bile salts** and chemically digested by **lipase**.
- **Bile** produced in the liver is stored in the gall bladder and is secreted into the duodenum via the bile duct.

- **Bile** contains:
  - **bile salts** that **emulsify** the lipids — mechanically break down large lipid globules into smaller droplets; this increases the surface area for the action of lipases
  - alkaline fluid containing $NaHCO_3$ to **neutralise** the stomach acid, providing an **optimum pH** for lipase
- **Lipases**, found in pancreatic juice, hydrolyse lipids into monoglycerides, fatty acids and glycerol.

## Chemical digestion of proteins

- **Endopeptidases** hydrolyse peptide bonds within the polypeptide chain to produce shorter polypeptide chains (peptides). Endopeptidases are found in gastric juice (for example, pepsin — optimum pH 2) and pancreatic juice (for example, trypsin — optimum pH 8).

The cells of the gut wall contain proteins, so endopeptidases are produced in an inactive form. Gastric glands secrete pepsinogen into the lumen of the stomach. The hydrochloric acid, which is secreted by different cells, converts the inactive pepsinogen into active pepsin. The pancreas secretes inactive chymotrypsin, which is converted to active trypsin by enterokinase in the lumen of the duodenum.

- **Exopeptidases** hydrolyse the terminal peptide bonds at the ends of the polypeptide chain to produce dipeptides and **amino acids**.
- Carboxypeptidases hydrolyse the terminal peptide bond at the carboxyl end of the polypeptide chain.
- Aminopeptidases hydrolyse the terminal peptide bond at the amine end of the polypeptide chain.
- Exopeptidases are produced in the pancreas and secreted into the duodenum (Figure 57). Dipeptidases are located on the cell surface membranes and within the epithelial cells of the small intestine. Some dipeptides are taken into the cell and digested intracellularly.

**Exam tip**

Many students think that bile salts are enzymes and that they hydrolyse lipids. This is wrong — bile salts only emulsify lipids, they do not chemically alter them.

**Knowledge check 34**

Explain the advantage of releasing both endopeptidases and exopeptidases into the gut.

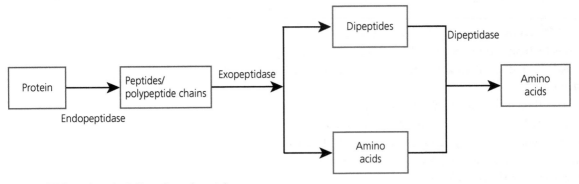

**Figure 57** The chemical digestion of proteins

# Summary

Table 13 summarises the chemical digestion of starch, proteins and lipids.

| | Carbohydrates | Triglycerides | Proteins | |
|---|---|---|---|---|
| **Mouth (pH 6.5–7.5)** | starch<br>↓ amylase<br>maltose | | | |
| **Stomach (pH 2)** | | proteins<br>↓ endopeptidase<br>↓ e.g. rennin<br>peptides | | |
| **Duodenum (pH 7–8.5)** | disaccharides<br>↓ carbohydrase*<br>monosaccharide | triglycerides<br>↓ lipase*<br>fatty acids and glycerol | proteins<br>↓ endopeptidase*,<br>↓ e.g. trypsin<br>polypeptides | polypeptides<br>↓ exopeptidase*<br>amino acids |

\* Produced by the pancreas

**Table 13** Summary of the chemical digestion of starch, proteins and lipids

# Absorption

Absorption takes place by **simple diffusion**, **facilitated diffusion**, **active transport** and **osmosis**. It takes place in the **ileum**, which has several adaptations:

- a large surface area:
  - very long — about 4 m in length
  - it is highly folded
  - the mucosa form finger-like villi (Figure 58)
  - the epithelial cells of the villi possess microvilli
- a short diffusion pathway — the epithelium is only one cell thick
- a steep diffusion gradient. Within each villus there are:
  - **blood capillaries**, which remove glucose and amino acids, keeping their concentration low
  - **lacteals** (part of the lymphatic system), which remove fatty acids and glycerol (and monoglycerides), keeping their concentration low

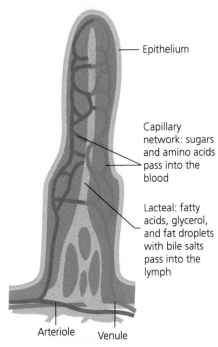

Figure 58 Cross section through a villus

Labels:
- Epithelium
- Capillary network: sugars and amino acids pass into the blood
- Lacteal: fatty acids, glycerol, and fat droplets with bile salts pass into the lymph
- Arteriole
- Venule

## Simple diffusion

Monoglycerides, fatty acids and glycerol are **non-polar molecules**, and diffuse through the phospholipid bilayer of the epithelial cells. Once inside the cells they are reformed into triglycerides. These triglycerides are then coated with a protein to form soluble structures called chylomicrons. They are then absorbed into the lacteals, and eventually pass into the blood via the thoracic duct.

## Facilitated diffusion and active transport

Some monomers, such as glucose and amino acids, are **polar** molecules, and they are transported from the gut through specific **carrier proteins** in the plasma membrane of the epithelial cells. They are then absorbed into the blood capillaries and transported via the hepatic portal vein to the liver.

- The epithelial cells (Figure 59) are adapted by having:
  - **microvilli** — which provide a larger surface area for absorption
  - many **mitochondria**, to produce the ATP required for active transport

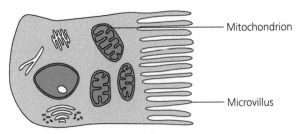

Labels: Mitochondrion, Microvillus

Figure 59 An epithelial cell from the ileum

> **Knowledge check 35**
>
> State one other function of the lymphatic system that you have studied in this unit.

The transport of glucose and amino acids from the lumen of the gut is dependent on the co-transport of sodium ions (see Figure 60). A sodium–potassium pump transports $Na^+$ out of the epithelial cell by active transport (1). This reduces the concentration of $Na^+$ inside the epithelial cell, generating a concentration gradient between the lumen of the gut and the cytoplasm of the epithelial cell. As a result, $Na^+$ and glucose (or a specific amino acid) bind to the carrier protein and enter the cell by facilitated diffusion (2). The concentration of glucose (and amino acids) inside the cytoplasm of the epithelial cell increases so that they pass into the blood via facilitated diffusion (3).

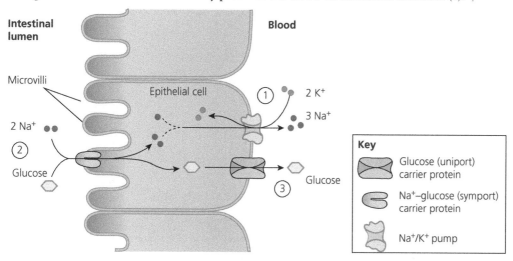

**Figure 60**

Facilitated diffusion is a relatively slow process, so some of these molecules are absorbed by active transport. This ensures that all the glucose and amino acids are absorbed from the lumen of the gut.

## Osmosis

Due to the absorption of solutes the water potential of the blood falls. This generates a **water potential gradient**, which causes a large volume of water to be absorbed into the blood by **osmosis**.

## The large intestine (colon)

The colon absorbs the remaining water, together with vitamins (secreted by microorganisms in the colon) in order to produce solidified faeces. Residues of undigested cellulose, bacteria and sloughed cells pass along the colon to be egested as faeces. Cellulose fibre is required to provide bulk and stimulate peristalsis.

# Carnivores and herbivores

Different mammals have evolved different adaptations to their guts to reflect their specialised diets.

## Carnivores

The gut of a carnivore is relatively short — this is due to the fact that protein is relatively easily digested. The dentition (Table 14) and powerful jaw muscles of a carnivore are adapted for catching and killing prey, cutting and tearing meat and

**Exam tip**

You will be expected to know the digestion of starch, proteins and lipids, the products formed and how these are absorbed. You will have already gained most of this knowledge from Unit 1, when you studied biological molecules and plasma membranes.

crushing bone (Figure 61). The jaw moves in a vertical plane, enabling the carnivore to open its mouth wide for catching and killing prey.

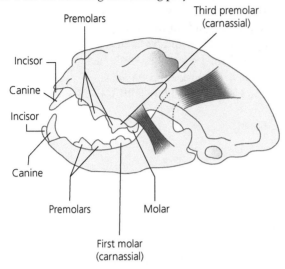

**Figure 61** A carnivore skull

| Teeth | Structure | Function |
|---|---|---|
| Incisors | Sharp | Strip flesh from the bone |
| Canines | Large, pointed and backward facing | Seize and kill prey |
| Carnassials (modified pre-molars and molars) | Modified premolars and molars with sharp cutting edges | Slice meat |
| Molars | Flattened with sharp edges | Crushing bone |

**Table 14** How the dentition of a carnivore is adapted to its diet

## Herbivores

The gut of a herbivore is relatively long — this is due to the fact that plant material, which contains cellulose, is difficult to digest. The dentition of a herbivore (Table 15) is adapted for cutting and grinding tough plant material (Figure 62). A grazing herbivore, such as a cow or sheep, has a jaw that moves in a horizontal plane, which produces a circular grinding action. A herbivore's teeth continue to grow throughout their lifetime (as they have open, unrestricted roots).

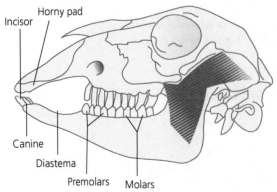

**Figure 62** A herbivore skull

**Knowledge check 36**

Prepare a table to compare the dentition of a fox and a sheep.

| Teeth | Structure | Function |
|---|---|---|
| Incisors | Small and flat topped — found on lower jaw only | Cut grass against a horny pad on the upper jaw |
| Canines | Indistinguishable from the incisors | |
| Diastema | Gap that separates the incisors from the premolars | Allows the manipulation of food by the tongue and to keep the freshly cropped grass separate from the cud |
| Premolars and molars | Large surface area, interlocking surfaces have sharp enamel ridges. | Efficient for grinding plant material |

**Table 15** How the dentition of a herbivore is adapted to its diet

## Ruminants

Most of a herbivore's diet consists of **cellulose**. However, mammals do not produce the enzyme **cellulase** to break down the cellulose. Many herbivores have therefore evolved **mutualistic relationships** with **gut bacteria** that can digest cellulose.

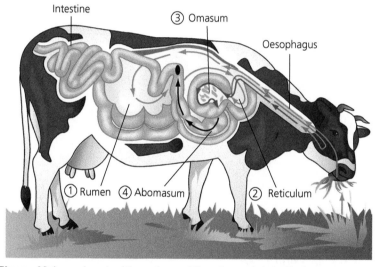

**Figure 63** A ruminant with regions of the stomach labelled

Ruminants (like cows and sheep) have a **four-chambered stomach** (Figure 63). The **rumen** is the first chamber and contains the mutualistic bacteria. The bacteria secrete cellulase, which hydrolyses the cellulose into glucose.

Ruminants 'chew the cud' that is, they carry out **reverse peristalsis** and bring food back into the mouth. This helps to further increase the surface area of the food but also mixes it with urea (found in their saliva). The urea provides a source of nitrogen for the bacteria so that they can synthesise amino acids and proteins.

Eventually the partly digested grass and dead bacteria are passed into the true stomach (abomasum) where protein digestion occurs. The digested food then passes to the small intestine where the soluble products of digestion are absorbed.

**Exam tip**

Make sure that you know the names of the different types of teeth and can link them to their mode of action.

**Exam tip**

A common mistake in exams is to state that cows have four stomachs. This is incorrect — they have one stomach that contains four different chambers.

**Knowledge check 37**

Explain why the rumen must be kept separate from the true stomach.

# Saprophytes and parasites

## Saprophytes/saprobionts

Saprophytes (also known as saprobionts) include **fungi** and some species of **bacteria** that feed on dead organic matter, that is, dead organisms and animal waste (faeces and urine). They carry out **extracellular digestion**:

- They secrete enzymes onto the dead organic matter.
- The enzymes hydrolyse the bonds in the organic molecules to produce small, soluble molecules.
- The small, soluble molecules are absorbed into the organism by diffusion and active transport.

## Parasites

Parasites can be classified as:

- Ectoparasites, which live on the body of the host, for example, the human body louse (*Pediculus humanus*).
- Endoparasites, which live inside the body of the host, for example, the pork tapeworm (*Taenia solium*).

### The head louse (*Pediculus humanus*)

The head louse feeds by sucking blood from the scalp of the host. It has claws to hold onto the hairs and lays eggs, which are glued to the base of the hair. Transfer between hosts is by direct contact.

### The pork tapeworm (*Taenia solium*)

The pork tapeworm (Figure 64) is an example of an endoparasite. Humans are its primary host and pigs are its secondary host. Humans get tapeworms by eating infected, undercooked pork. The pig then becomes infected by ingesting human faeces (from untreated sewage) that contains tapeworm eggs.

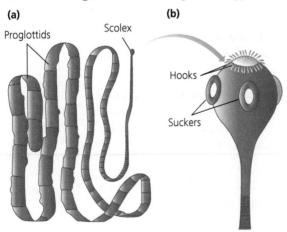

**Figure 64** A pork tapeworm: (a) whole; (b) close-up of scolex

A **saprophyte** is an organism that secretes enzymes onto dead organic matter, carries out extracellular digestion and then absorbs the soluble products by diffusion.

**Exam tip**

Fungi and bacteria secrete enzymes, they do not excrete them — excretion is removal of the waste products of cell metabolism.

A **parasite** is an organism that lives in or on another organism, its host, obtaining nourishment from and causing harm to the host.

*Adaptations*

- Its head (scolex) has curved hooks and suckers for attachment to the gut wall.
- It has a thick cuticle and produces inhibitory substances on its body surface, which protect it from the host's immune responses and prevent its digestion by the host's enzymes.
- Its long, thin body provides a large surface area-to-volume ratio for the absorption of pre-digested food by diffusion and active transport.
- It is a hermaphrodite and reproduces sexually by self-fertilisation. Each segment contains both male and female reproductive organs and can produce large numbers of embryos/eggs.
- The eggs have a resistant shell, which enables them to survive outside their host until they are eaten by a pig.

**Knowledge check 38**

State two ways in which infection by pork tapeworm can be prevented.

**Exam tip**

The fact that tapeworms are hermaphrodites confuses some students and they make incorrect statements such as:

- 'They reproduce asexually' — they do not; tapeworms produce both male and female gametes and are capable of self-fertilisation.
- 'They produce many eggs to increase the chances of fertilisation' — this is incorrect as the eggs are already fertilised. They produce many eggs (which contain embryos) to increase the chances of reproductive success and the survival of the species.

**Practical work**

During this topic you will have made observations using microscope slides of sections of the duodenum and the ileum. In the exam you may be given photomicrographs or drawings of these tissues. It is important that you can identify the various structures and link them to their functions.

You will also have examined specimens and slides of the pork tapeworm. It is important that you can link the structural adaptations of the parasite to its mode of nutrition.

## Summary

After studying this topic you should be able to demonstrate and apply your knowledge and understanding of:

- autotrophic organisms, which may be photoautotrophic or chemoautotrophic
- heterotrophic organisms, which may be holozoic, saprotrophic/saprobiotic or parasitic
- nutrition in unicellular organisms, for example, amoeba, which carry out intracellular digestion
- holozoic digestion, which involves the digestion of food substances in an internal gut. The structure of the gut may range from a simple, undifferentiated, sac-like gut with a single opening, such as is present in *Hydra*, to a tube gut that has different openings for ingestion and egestion and specialised regions for the digestion of different food substances

- the human gut, which is adapted to a mixed, omnivorous diet, and which requires different enzymes and different conditions along its length to digest and absorb different food types and molecules
- the adaptations that are shown by herbivores, and in particular ruminants, to a high cellulose diet, and the adaptations that are shown by carnivores to a high protein diet, including adaptations of dentition and regions of the gut
- saprotrophic nutrition of organisms including fungi, which involves extracellular digestion of food substances followed by absorption of the products of digestion into the organism
- the highly specialised adaptations shown by parasites that enable them to obtain their nutrition at the expense of a host organism

# Questions & Answers

## The unit test

There are 80 marks available in the Unit 2 exam. The majority of these marks are from a range of short and longer structured questions; 9 marks are available from an extended response question that will assess the quality of written communication as well as the content of the answer.

When exam papers are being prepared, the examiner must try to ensure that all the topics covered in the unit are assessed, so you should expect to get questions from each of the four topic areas in Unit 2. However, Unit 2 covers a lot of material, so it would be impossible to be asked a question on everything. Examiners must also set questions that test the specific assessment objectives. You may find it useful to understand the weighting of the assessment objectives that will be used.

The assessment objectives are weighted as follows:

| Assessment objective | Brief summary | Approximate percentage of marks available | Approximate number of marks available |
|---|---|---|---|
| AO1 | Demonstrate knowledge and understanding of scientific ideas, processes, techniques and procedures | 35 | 28 |
| AO2 | Apply knowledge and understanding of scientific ideas, processes, techniques and procedures | 45 | 36 |
| AO3 | Analyse, interpret and evaluate scientific information, ideas and evidence, including in relation to issues, to make judgments and reach conclusions, and to develop and refine practical design and procedures | 20 | 16 |
| | Total | 100 | 80 |

Approximately half of the marks available target AO2 (apply knowledge and understanding of scientific ideas, processes, techniques and procedures), so many of the questions will be written in an unfamiliar context. However, it is important for you to understand that you should have learned all of the biology to enable you to answer these questions. First you need to recognise which part of the specification the question is targeting, and then you can apply your knowledge to the particular scenario that has been presented. You may find it useful to use a highlighter pen to select the important information provided in the question, as this will help you when you have to refer back to the stem of the question.

A minimum of 8 marks (10%) will relate to the assessment of your mathematical skills, and a minimum of 12 marks (15%) will relate to the assessment of your practical skills. It is therefore important that you review your practical work and practise doing calculations as part of your revision, to ensure that you gain the marks on these questions.

## About this section

This section contains questions on each of the topics. They are written in the style of the questions in Unit 2, so they will give you an idea of what you will be asked to do in the exam. Each question is followed by tips on what you need to do to gain full marks (shown by the icon ⓔ). The student responses are also followed by comments. These are preceded by the icon ⓔ and highlight where credit is due. The comments also point out areas for improvement in the weaker answers, and highlight specific problems and common errors such as lack of clarity, irrelevance, misinterpretation of the question and mistaken meanings of terms.

Each question is attempted by two students, student A (a weaker answer) and student B (a strong answer). The students' answers, along with the comments, should help you to see what you need to do to score a good mark — and how you can easily not score a good mark even though you probably understand the biology.

## Question 1 The evolutionary relationships between organisms

Giant pandas (*Ailurupoda melanoleuca*) inhabit the remote bamboo forests of the Himalayas and south China. Although they belong to the order Carnivora, they have an herbivorous diet, and feed almost exclusively on the leaves and shoots of bamboo. The giant panda, like all other mammals, possesses pentadactyl limbs. They appear to have six digits on their front paws but one is not a true digit. As shown in Figure 1, it is an elongated wrist bone that is equipped with muscles to hold and strip bark from bamboo.

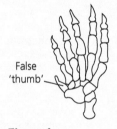

False 'thumb'

**Figure 1**

(a) Explain how natural selection could have brought about the evolution of the panda's 'thumb'.

(5 marks)

(b) Since its 'discovery' in 1862, western scientists have been debating which family the giant panda belongs to. Some scientists thought that they were closely related to red pandas (*Ailurus fulgens*), which also inhabit the bamboo forests of the Himalayas and South China. At that time, the red panda belonged to the family Procyonidae, which also includes racoons. Other scientists thought that the giant pandas were more closely related to bears, which belong to the family Ursidae.

Table 1 compares several characteristics shown by the giant panda, the American black bear and the red panda. This morphological and behavioural data was used in the classification of the giant panda prior to 1985.

| Characteristic | Giant panda | American black bear (family Ursidae) | Red panda (family Procyonidae) |
|---|---|---|---|
| Height or length/cm | 120–180 | 120–300 | 30–70 |
| Mass/kg | 75–120 | 27–450 | 0.8–18 |
| Dentition and diet | | | |
| Number of teeth | 40 | 42 | 40 |
| Number of molar teeth | 8 | 10 | 8 |
| Diet | Herbivorous | Omnivorous | Herbivorous |
| Behaviour | | | |
| Hibernates | No | Yes | No |
| Scent marks territory | Yes | No | Yes |
| Vocalisations | Yes | No | Yes |

Table 1

Phylogenetic trees showing possible evolutionary relationships between the three species are shown in Figure 2.

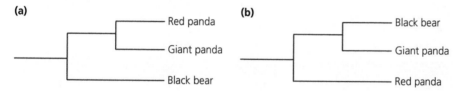

(a)
- Red panda
- Giant panda
- Black bear

(b)
- Black bear
- Giant panda
- Red panda

Figure 2

(i) The phylogenetic tree in Figure 2(a) suggests that the giant panda is more closely related to the red panda than the black bear. What evidence did they use to justify this conclusion? (2 marks)

(ii) In 1985 DNA analysis was used to solve the classification problem of the giant panda. The evidence was used to produce the phylogenetic tree in Figure 2(b). Use your knowledge of DNA analysis to explain how the scientists came to this conclusion. (3 marks)

(iii) What conclusions can be drawn about the evolution of dentition in the giant panda and in the red panda? (1 mark)

(c) Complete the table below for the classification of the giant panda based upon the DNA evidence. (2 marks)

| Kingdom | Animalia |
|---|---|
| Phylum | Chordata |
| | Mammalia |
| | Carnivora |
| Family | |
| Genus | |
| Species | |

Total 13 marks

e In this question you have been provided with a lot of information about giant pandas. It is important that you take your time, and read each part of the question carefully. The question is quite straightforward and is simply placing natural selection and classification in a particular context. For this reason, the majority of the marks test your ability to apply knowledge and understanding (AO2). Part (b)(iii) requires you to use information in the text and in the table to draw a conclusion; therefore it is testing your ability to evaluate scientific information (AO3).

---

**Student A**

**(a)** Pandas may have started off without a thumb but a mutation **a** may have occurred to make this false thumb. The pandas without the thumb that were less well adapted may have died out and left the pandas with the false thumb to reproduce **b**.

**(b) (i)** The giant panda's dentition and diet, as well as its behaviour, shows that it is closely related to red pandas **c**.

**(ii)** DNA analysis would show that the two most closely related organisms have the most similar banding pattern in their DNA. Therefore the giant panda's DNA would be most similar to that of the black bear **d**.

**(iii)** They both have the same number of teeth and molar teeth, so the dentition of the red panda and the giant panda have evolved to be more similar to each other than to the dentition of the black bear **e**.

**(c)** Class and order **f**; Procyonidae, Ailurupoda and melanoleuca **g**.

---

e **5/13 marks awarded** **a** The student has correctly stated that the 'thumb' would have arisen by mutation and **b** demonstrates a basic understanding that this would lead to differential survival and reproduction. However, the answer lacks detail and therefore only gains 2/5 marks available. **c** The student has chosen the correct characteristics but once again the answer lacks detail, so only 1 mark is awarded. **d** The student demonstrated a knowledge of DNA profiling but has failed to fully apply it to the scenario presented, so 1 mark is awarded. **e** Incorrect, the student has not read the question carefully and has simply quoted data from the table. **f** Correct. **g** Again the student has failed to read the question carefully and has stated the wrong family for the giant panda.

**Student B**

**(a)** The offspring of giant pandas have varied phenotypes. In this case some offspring had an elongated wrist bone [a]. The offspring with this variation were better adapted to the bamboo forests, as they were able to hold and strip the leaves from bamboo more efficiently than pandas without the elongated wrist bone [b]. This beneficial characteristic gave these pandas a selective advantage, and therefore a better chance of survival over the pandas without the beneficial characteristic [c]. These pandas will reproduce and their offspring will also possess the elongated wrist bone [d]. Over successive generations the proportion of pandas with the elongated wrist bone in the population will increase [e].

**(b) (i)** Giant pandas and red pandas both have 40 teeth, of which 8 are molars, as well as both being herbivorous [f]. The American black bear has 42 teeth, of which 10 are molars.

**(ii)** As different species diverge from their common ancestor they will accumulate differences in their DNA base sequences. By analysing the base sequences of all three species [g] they would have found fewer differences between the base sequence of the giant panda and the black bear [h] suggesting that they shared a more recent common ancestor [i].

**(iii)** The dentition of the giant panda and the red panda must be due to convergent evolution [j].

**(c)** Class and order [k] and Ursidae, Ailurupoda and melanoleuca [l].

**e** **12/13 marks awarded** The student has given a detailed account and gains nearly all of the marks available. The answer makes reference to variation [a], selective advantage [b], differential survival and reproduction [c], inheritance [d] and the effect on the population [e]. [f] The student has provided the correct detail regarding anatomy but has failed to make reference to common behaviours; they gain 1 mark. The student has demonstrated a knowledge of DNA sequencing and has fully applied it to the scenario presented, including a comparison of all three species [g], reference to the results [h], and a comment about their common ancestor [i]; all 3 marks awarded. [j], [k] and [l] correct.

**e** **Although there is a lot of information in this question, the biological knowledge being assessed is quite straightforward. If you are well prepared and read the questions carefully, you should be able to gain most of the marks available on these types of questions. Student A has given answers that lack the detail and has failed to apply their knowledge to the situation given. Student A is awarded 5 marks (grade U). In contrast, student B is awarded 12 marks (grade A) for clearly demonstrating the ability to recall and apply their knowledge.**

# Question 2 Adaptations for gas exchange

(a) Figure 3 shows an amoeba (*Amoeba proteus*) and an earthworm (*Lumbricus terrestris*). Amoebas vary in size from 250 µm to 750 µm, while *Lumbricus terrestris* varies in size from 110 mm to 200 mm long and from 7 mm to 10 mm in diameter. Amoebas and earthworms both carry out gas exchange across their body surface.

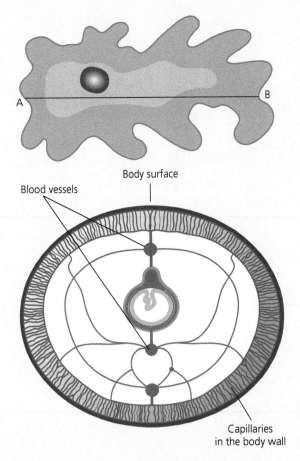

Figure 3

(i) Describe the four properties that are common to the gas exchange surfaces of both amoebas and earthworms. (2 marks)

(ii) The magnification of the photomicrograph used to produce the diagram of the amoeba was ×140. Calculate the actual size, in micrometres (µm), of the amoeba from point A to point B, and give your answer to three significant figures. (2 marks)

(b) Table 2 shows the relationship between the diffusion distance and the approximate time required for the diffusion of oxygen. Use the information to explain why the earthworm requires a transport system but the amoeba doesn't. (5 marks)

| Diffusion distance/mm | Approximate time required for the diffusion of $O_2$/seconds |
|---|---|
| 0.001 | $2.38 \times 10^{-4}$ |
| 0.01 | $2.38 \times 10^{-2}$ |
| 0.1 | 2.38 |
| 1 | $2.38 \times 10^{2}$ |
| 10 | $2.38 \times 10^{4}$ |

Table 2

(c) Figure 4 represents the pattern of spiracles of an insect (legs and wings removed). Each spiracle has a valve, which can be closed or opened to control the flow of air into the body. The graphs show the opening and closing of the spiracles in the insect at rest and also how regular muscular movements expand and compress the abdomen.

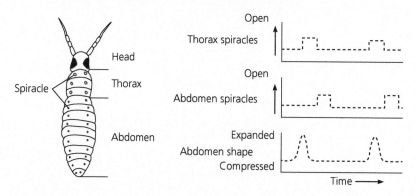

Figure 4

(i) State what is meant by the term ventilation. (1 mark)

(ii) Use the information provided to explain how the insect ventilates its tracheal system. (5 marks)

(iii) The graphs show that there are long periods when the spiracles are closed. Explain why this is important for the insect's survival. (1 mark)

Total 16 marks

ⓔ All three assessment objectives are tested in this question. Parts (a)(i) and (c)(i) simply require demonstration of knowledge and understanding, whereas parts (b) and (c) require you to apply your knowledge and understanding of gas exchange. The question also presents data in different formats; your ability to analyse this information and to reach conclusions tests AO3. Part (a)(ii) assesses your mathematical skills — the ability to calculate actual size and magnification from photomicrographs can be assessed in both Unit 1 and Unit 2.

**(a) (i)** They are both permeable to gases and have a large surface area and a short diffusion pathway a.

**(ii)** 224 µm b

**(b)** The table shows that as the diffusion distance increases the time taken for diffusion of oxygen increases proportionally. Amoeba vary in size from 250–750 µm whereas earthworms vary from 110–220 mm in length. Therefore the *Amoeba* has a very large surface area to volume ratio in comparison to the earthworm c. The earthworm has blood vessels close to the surface of the skin to provide a short diffusion pathway d. The earthworm is also larger and therefore has a greater requirement for oxygen than the amoeba.

**(c) (i)** Ventilation is the movement of the respiratory medium over the respiratory surface e.

**(ii)** The thoracic spiracles open before abdomen spiracles. Prior to the abdomen spiracles opening the abdomen expands. This causes the volume to increase and pressure inside the insect to decrease f. When the spiracles open air moves into the branched tracheal system g.

**(iii)** When the spiracles are open the insect has increased water loss h.

⊖ **4/16 marks awarded** a The student has correctly stated three properties and not four, so gains 1 mark. b The answer is incorrect and because the student has failed to show how they arrived at the answer, they lose both marks. c The student has made reference to the data relevant to the question, unfortunately they have then made reference to surface area to volume ratio and not length of the diffusion pathway. However they have made a correct reference to this in relation to the earthworm d, and gain 1 mark. e Correct. In (c)(ii) the student has attempted to interpret the data and has given a correct statement about the relationship between volume and pressure f but then only gives a vague explanation for the movement of air into the insect g. h The student has recalled the correct biology but hasn't actually answered the question.

**(a) (i)** Both amoeba and earthworm have a gas exchange surface with a large surface area, short diffusion pathway. Both are also permeable to gases and moist a.

**(ii)** A-B = 64 mm    therefore 64 mm/140 = 0.457 mm = 457 µm b

**(b)** The amoeba does not need a transport system as its maximum size is 750 µm meaning it would take between 2.38 and 238 seconds for molecules of oxygen to diffuse from the environment to the centre of the amoeba (maximum radius 375 µm) c. Diffusion alone is sufficient to meet its demand for oxygen d. The earthworm however has a maximum radius of 5 mm, therefore it would take between 234 seconds (4 minutes) and 397 minutes for oxygen to diffuse into the centre of the earthworm, therefore diffusion alone will not meet its demand for oxygen █. The earthworm's blood vessels are close to its body surface so provides a short diffusion pathway for oxygen to diffuse quickly into the blood █. The oxygenated blood can then be transported around the body at a faster rate than diffusion █.

**(c) (i)** Ventilation is the movement of the respiratory medium over the respiratory surface █.

**(ii)** Firstly the abdomen expands; this will increase its volume and decrease the pressure inside the abdomen █. The thoracic spiracles then open and air will enter the tracheal system down a pressure gradient █. The abdomen then compresses, which reduces the volume and increases the pressure inside the abdomen █. The thoracic spiracles close and the abdominal spiracles open █. This shows a one-way flow of air through the tracheal system; it enters through the thoracic spiracles and exits through the abdominal spiracles █.

**(iii)** This reduces water loss and helps to prevent the earthworm drying out █.

ⓔ **15/16 marks awarded** ⓐ The student has correctly stated four properties and gains both marks. ⓑ The answer is incorrect, but the working is correct so they gain 1 mark. ⓒ and ⓓ The student has used the data correctly to explain why the amoeba doesn't need a transport system; ⓔ ⓕ and ⓖ and has used both the data and the diagram of the earthworm to give a complete answer to the question. ⓗ Correct. ⓘ to ⓜ The student has given a well-structured answer clearly linking the data provided to their knowledge of ventilation. ⓝ Correct.

ⓔ **Diffusion is only efficient over very short distances, and it is this concept that is being assessed in this question. The ability to apply your knowledge to help interpret data is an important skill in science. Student A demonstrates a basic understanding of gas exchange and the marks awarded are for recall, but not for application. They gain 4 marks (grade U). In contrast, student B gains 15 marks (grade A) for clearly demonstrating the ability to apply their knowledge in an unfamiliar context.**

## Question 3 Adaptations for transport in plants

**(a)** A student carried out an investigation on the stomatal density in the leaves of an English oak tree. She used nail varnish to prepare an impression of the lower epidermis of the leaf. She observed the impression under the microscope and counted 42 stomata. She then measured the diameter of the field of view as 435 μm.

**(i)** Calculate the mean stomatal density per mm$^2$ and give your answer to three significant figures. ($\pi = 3.14$) (3 marks)

**(ii)** Explain how the practical could be modified to ensure that the student's results were more reliable. (1 mark)

**(b)** Explain why guard cells are important to the functioning of the oak tree. (2 marks)

**(c)** The leaves of pineapple plants (*Ananas comosus*) tend to have a stomatal density of 80 mm$^{-2}$. At night the leaves take in $CO_2$, which joins with a three-carbon compound to form malic acid (a four-carbon compound). The malic acid accumulates in the central vacuole of the cells during the night. During the day the accumulated malic acid leaves the vacuole and is broken down to release $CO_2$, which can then be used in photosynthesis.

What does the information given suggest about the environment in which pineapples grow? Explain how you arrived at your conclusion. (3 marks)

**Total 9 marks**

ⓔ Investigating the stomatal density of a leaf is a specified practical and the examiner will therefore make the assumption that you can recall information about the procedure. In part (a) your mathematical skills are being tested (AO2) as well as your ability to refine practical design (AO3). Part (b) is recall of information (AO1) while part (c) tests your ability to apply your knowledge and understanding of scientific ideas (AO2) and to draw a conclusion (AO3).

---

**Student A**

**(a) (i)** area of field of view = 3.14 × 0.435$^2$ = 3.14 × 0.189225 = 0.594 mm$^2$ ⓐ
mean stomatal density = 0.594/42 ⓑ = 0.0141/mm$^2$

**(ii)** She could get more reliable results by counting the number of stomata in different fields of view so that a mean could be calculated ⓒ.

**(b)** They open to allow the entry of $CO_2$, which is necessary for photosynthesis to occur ⓓ. They also allow the entry of $O_2$ at night so the plant can carry out respiration ⓔ.

**(c)** Pineapples store the $CO_2$ gathered at night. This means they grow in an atmosphere where there is a low $CO_2$ concentration. This is concluded as to photosynthesise plants need $CO_2$ and the pineapple gets this from the $CO_2$ entering during the day and which they stored in the previous night. It suggests that the $CO_2$ absorbed during the day was not a sufficient amount to make enough glucose for the plant ⓕ.

---

ⓔ **2/9 marks awarded** ⓐ The student has incorrectly calculated the area of the field of view; they have made the common mistake of using the diameter given and not calculating the radius. They have also incorrectly divided the area by the number of stomata ⓑ and therefore the answer is incorrect; they gain no marks. ⓒ Correct. ⓓ The student gains 1 mark for making a statement referring to the opening of stomata for gas exchange linked to photosynthesis. However, the second statement ⓔ is a common misconception that plants only carry out respiration at night. ⓕ The student has used the information provided, but has made the wrong conclusion. They have failed to read the information carefully, and have assumed that the stomata are open both during the day and at night, which has led to the confusion — they gain no marks.

**Student B**

**(a) (i)** The radius of the field of view = 217.5 μm or 0.2175 mm
Area of field of view = 3.14 × 0.2175² = 0.149 mm²
Mean stomatal density = 42/0.149 = 282/mm² **a**

**(ii)** The experiment should be repeated at least three times, on different leaves, so that mean number of stomata could be calculated **b**.

**(b)** The guard cells control the opening and closing of the stomata. When open $CO_2$ can enter the leaves, enabling the oak tree to photosynthesise **c**. When closed, at night, the guard cells help to minimise water loss, as there is no light for photosynthesis **d**.

**(c)** The information suggests that the pineapple grows in an environment of limited water availability **e**. The stomatal density of pineapple leaves is much lower than the stomatal density of the oak leaves, suggesting that pineapple plants are xerophytes **f**. The pineapple only takes in $CO_2$ at night and then forms malic acid from which to form $CO_2$ during the day. This enables the pineapple to carry out photosynthesis during the day with its stomata closed, therefore reducing water lost **g**.

**e** **9/9 marks awarded** **a** The student has read the question carefully and converted the diameter given into the radius and converted from μm to mm. They have also correctly calculated the density and given their answer to three significant figures. **b** Correct. The student has made correct references to the role of guard cells in gas exchange **c** and the reduction in water loss **d**. The student has arrived at the correct conclusion **e** in part (c) and has used the information provided to explain how they arrived at their conclusion **f** and **g**.

**e** You must make sure that when you revise you go through the set practical activities that you have completed on the course and that you practise mathematical problems. Student A has made some common errors with the calculation and demonstrates a limited knowledge and understanding of the topic. They gain 3 marks (grade U). In contrast, student B demonstrates good subject knowledge and the ability to interpret the data provided. They have demonstrated good practical and mathematical skills and gain 9 marks (grade A).

## Question 4 Adaptations for transport in animals

Carbon dioxide and 2,3-diphosphoglycerate (2,3-DPG) are two molecules that affect the ability of haemoglobin to transport oxygen. Carbon dioxide is produced during aerobic respiration and 2,3-DPG is produced during anaerobic respiration in erythrocytes (red blood cells).

Figure 5 shows an erythrocyte.

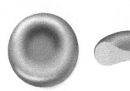

Top view     Side view (cut)

**Figure 5**

**(a)** Erythrocytes lack organelles such as a nucleus and mitochondria. Explain how the lack of organelles is important to the functioning of the cell. (2 marks)

Figure 6 shows three oxygen-haemoglobin dissociation curves.

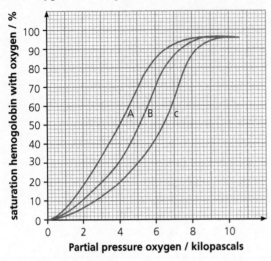

**Figure 6** Oxygen–haemoglobin dissociation curves.
A, adult Hb at pH 7.6; B, adult Hb at pH 7.4; C, adult
Hb in the presence of 2,3-DPG

**(b) (i)** Curve B represents the curve for normal adult haemoglobin at pH 7.4.
Explain the significance of the shape of this curve. (3 marks)

**(ii)** Explain the advantage, to the mountaineer, of an increase in 2,3-DPG
concentration. (2 marks)

**(c)** The following information was obtained from a website about
mountaineering. When you ascend to high altitude your body has to adapt to
the lack of oxygen. As a consequence you breathe faster and more deeply to
maximise the amount of oxygen that can get into the blood from the lungs.
This hyperventilation also results in more $CO_2$ being expelled from the lungs.
People who ascend too quickly are at risk of developing acute mountain
sickness. The symptoms of acute mountain sickness include headache,
nausea and lethargy (tiredness).

Use all the information provided to explain why people with acute mountain
sickness feel lethargic. (4 marks)

**Total 11 marks**

ⓔ Parts (a) and (b)(i) require you to recall information that you have learned on
the course and are therefore assessing AO1. Parts (b)(ii) and (c) assesses AO2 as
you are required to apply your knowledge of oxygen–haemoglobin dissociation
curves to this particular example. Part (c) requires you to analyse the information
from different parts of the question to reach a conclusion — this is AO3.

### Student A

(a) It means that the red blood cell has a biconcave disc shape giving it a larger surface area ⓐ.

(b) (i) Because the curve is S-shaped it means that it will load oxygen in the lungs because there is a high $pO_2$ in the lungs. And because there is a low $pO_2$ in the tissues it will unload oxygen ⓑ.

(ii) 2,3-DPG shifts the curve to the right ⓒ so that it will unload the oxygen faster ⓓ.

(c) If the person starts hyperventilating then they won't be getting enough oxygen into their bodies. This will make them feel light-headed. Because there is less oxygen at high altitude the light-headedness will give them a headache and make them feel sick and not want to carry on climbing ⓔ.

ⓔ **3/11 marks awarded** ⓐ The student has only made one correct statement and gains 1 mark. ⓑ The student has gained 1 mark for correctly stating where oxygen will be loaded and unloaded, but has failed to mention the effect of $pO_2$ on haemoglobin's affinity for oxygen or made reference to the steep part of the curve. ⓒ 1 mark is gained for stating that the curve will shift to the right, but then an incorrect statement has been made about the effect on haemoglobin ⓓ. ⓔ The student has tried to use their own knowledge to answer the question instead of using the information provided and fails to gain any marks.

### Student B

(a) This creates space within the cytoplasm for more haemoglobin so that it can transport more oxygen. The lack of nucleus creates a biconcave disc shape, which increases the surface area of the cell, which will increase the rate of diffusion of oxygen ⓐ.

(b) (i) At high partial pressures of oxygen, e.g. in the lungs, haemoglobin has a high affinity for oxygen and will load it to form oxyhaemoglobin. At low partial pressures of oxygen, e.g. in the tissues, haemoglobin has a low affinity for oxygen so will unload it ⓑ. In the middle part of the curve, a small drop in $pO_2$ results in a relatively large drop in percentage saturation, so more oxygen is unloaded to the tissues ⓒ.

(ii) 2,3-DPG will cause haemoglobin to have a lower affinity for oxygen ⓓ. Therefore more oxygen is unloaded to the tissues for aerobic respiration ⓔ.

(c) Hyperventilating will cause more $CO_2$ to be removed from the body and therefore the pH of the blood will increase ⓕ. The graph shows that this will increases haemoglobin's affinity for oxygen as it shifts the dissociation curve to the left ⓖ. Therefore less oxygen will be unloaded to the tissues and the mountaineer will feel lethargic ⓗ.

(e) **10/11 marks awarded.** (a) 2 marks awarded for two correct statements. (b) The student provides a concise explanation linking the effect of $pO_2$ on haemoglobin's affinity for oxygen and consequently where it will be loaded and unloaded, and has made reference to the steep part of the curve (c). The student has used the information to explain the effect on haemoglobin (d) and the advantage to the mountaineer (e). (f) The student has linked the information in the text to the graph and understands the relationship between $CO_2$ and pH. They understand the effect that this would have on haemoglobin's affinity for oxygen (g). Unfortunately, the last statement (h) simply repeats the stem of the question instead of making the link between less oxygen and less respiration, and therefore less ATP, for processes such as muscle contraction; they gain 3/4 marks available.

(e) **Many students have difficulty answering questions on dissociation curves. However, you can gain good marks if you keep in mind some basic principles: if the curve is to the left, haemoglobin has a higher affinity for $O_2$ and is better at loading $O_2$; if the curve is to the right, haemoglobin has a lower affinity for $O_2$ and is better at unloading $O_2$.**

**Student A has gained marks for simple recall of knowledge; however the majority of the question is assessing application of knowledge. The student has also failed to provide enough information for the marks available for the question — if a question is worth 2 marks you must give two 'pieces' of biological information. They gain 3 marks (grade U). In contrast, student B demonstrates a good understanding of the biology. They are well prepared, have clearly learned the biology and demonstrate the ability to apply this knowledge to unfamiliar situations. They gain 10 marks (grade A).**

## Question 5 Adaptations for nutrition

**Coeliac disease is believed to be due to a genetically determined abnormal immunological reaction to the protein gluten, which is present in wheat. The disease is a common digestive problem and leads to damage to the wall of the ileum, resulting in the degeneration of the villi. Symptoms of coeliac disease include the passage of watery stools, weight loss and fatigue in adults, and poor growth rates in children.**

**Use your knowledge of the structure and function of the ileum to explain how the symptoms of coeliac disease develop.**

(9 QER)

(e) At the end of every Unit 2 test there will be an extended response ('essay') question for which the quality of your extended response (QER) is assessed. The essay question will test both your ability to demonstrate (AO1) and apply (AO2) your knowledge and understanding of biology. The question also assesses the quality of your written communication. The mark scheme contains three bands (7–9 marks, 4–6 marks and 1–3 marks) and each band contains two statements, one relating to the biological content and the other relating to the communication of your answer. If you have learned your biology and answer all parts of the question then your answer is likely to be in the top band.

**Student A**

If a person has coeliac disease then their villi are damaged. villi are important in the absorption of the food. Villi have a large surface area to make absorption more efficient so if they are damaged they are not as good at absorbing food [a]. If less food is being absorbed then the person will lack energy [b]. because of this the person will start 'burning fat' and this will cause them to loose weight [c]. If children can't absorb enough food then they won't grow as well as they would if they absorbed more food [d]. A person with coeliac disease will produce watery stools because of osmosis. If food like glucose isn't absorbed then it will mean that the water potential in the gut is low. water moves by osmosis from a high water potential to a low water potential so water will move into the gut and make to stools watery [e].

[e] **4/9 marks awarded.** The student has attempted to answer all parts of the question. They have made reference to the role of the villi in absorption [a] and noted that damage will lead to the symptoms of fatigue [b], weight loss [c], poor growth [d] and the production of watery stools [e]. However, there is no reference to the role of the ileum in digestion. The majority of the answer lacks detail required at this level — this places the answer in the middle (4–6 marks) band. The quality of the written communication is poor; the answer is presented as a single paragraph and many of the sentences do not begin with a capital letter. There is also a lack of technical terminology throughout as well as spelling mistakes. This places the answer at the lower end of the band.

**Student B**

If a person suffers from coeliac disease then the degeneration of the villi in the ileum reduces their surface area. The villi are important as the final stages of digestion occur on the epithelial cells and the absorption of the soluble products of digestion occurs in the ileum. This will result in less efficient digestion and absorption of the soluble products of digestion, which can explain the symptoms of the disease [a].

If less food absorbed then solutes, such as glucose, will remain in the lumen of the gut. Therefore the water potential in the lumen of the gut will also remain low. Therefore less water is absorbed from the gut by osmosis, resulting in the passage of watery stools [b].

Maltose is hydrolysed into glucose on the membranes of the epithelial cells of the villi. If the villi are damaged then less glucose is absorbed into the blood. With less glucose being absorbed into the blood the person will have less energy (fatigue) as glucose is required for respiration [c]. If the person can't absorb enough glucose for respiration then they will start to break down lipids which will lead to weight loss [d].

Dipeptides are hydrolysed into amino acids inside the epithelial cells of the villi. Damage to the villi will mean that less amino acids are being absorbed into the blood. A lack of amino acids leads to a reduction in protein synthesis. This will result in poor growth rates in children [e].

(e) **9/9 marks awarded.** The student has provided a detailed answer to the question; they have also structured their answer to provide a concise explanation to different parts of the question. They have made reference to the role of the villi in both digestion absorption a, and that damage will lead to the production of watery stools b, symptoms of fatigue c, weight loss d and poor growth rates in children e. With each explanation the student demonstrates detailed biological knowledge and the ability to apply this knowledge to the scenario given — this places the answer in the top (7–9) band. The quality of the written communication is excellent; the answer is well structured and logically presented, with detailed biological terminology used throughout. This places the answer at the upper end of the band.

(e) **Although the question is set in an unfamiliar context, the biological knowledge being assessed is straightforward. The question states the biology that is required (the structure and function of the ileum) to answer the question in a particular context (symptoms of coeliac disease). If you are well prepared, and read the questions carefully, you should be able to gain most of the marks available on these types of questions. Student A has attempted to answer all parts of the question but the answer shows a limited knowledge of the functions of the ileum and a lack of the detail expected at this level. S/he gains 4 marks (grade E). In contrast student B has given a detailed account clearly demonstrating their ability to recall and apply their knowledge of digestion and absorption. S/he is well prepared and has spent time learning the biology required and gains 9 marks (grade A).**

## Knowledge check answers

**1** The pool, as it has more species present.

**2**

| Species | Number ($n$) | ($n - 1$) | $n(n - 1)$ |
|---|---|---|---|
| Minnow | 12 | 11 | 132 |
| Stonefly larva | 16 | 15 | 240 |
| Mayfly larva | 19 | 18 | 342 |
| Dragonfly larva | 9 | 8 | 72 |
| Freshwater shrimp | 22 | 21 | 462 |
| Bloodworm | 0 | 0 | 0 |
| Total ($N$) | 78 | $\Sigma$ | 1248 |
| $N - 1 =$ | 77 | | |
| $N(N - 1) =$ | 6006 | | |

$D = 1 - (1248/6006)$
$D = 1 - 0.21$
$D = 0.79$

**3** Humans are *Homo sapiens* and the grey wolf is *Canis lupus*.

**4** Ground finches and tree finches are the most closely related as they share the most recent common ancestor, i.e., the two groups diverged most recently.

**5** False; they look similar, however they are analogous structures and not homologous structures. (Note: it is important that you do not confuse analogous and homologous structures.)

**6** Plants = cellulose; fungi = chitin; prokaryotes = peptidoglycan (murein)

**7** They are permeable to gases, thin (to provide a short diffusion pathway) and they have a large surface area. (Note: they are also moist but this will not gain credit in questions about fish gills or other aquatic organisms! Not all gas exchange surfaces have ventilating mechanisms or an associated blood supply — e.g. insects.)

**8** Internal lungs help to reduce water loss and heat loss from the body.

**9** Ventilation maintains a steep $O_2$ concentration gradient between the environment and the organism, and increases the rate of gas exchange. This provides more $O_2$ for an increased rate of aerobic respiration.

**10** There are many alveoli, providing a large surface area. The walls of the alveoli are thin, which provides a short diffusion pathway for oxygen to enter the blood. The numerous blood vessels show that there is a good blood supply to transport oxygen away from the alveoli. This is important to help maintain a steep oxygen concentration gradient between the alveoli and the blood.

**11** To maintain a constant body temperature they have very high metabolic rates, as aerobic respiration releases heat energy; they therefore require large volumes of oxygen.

**12** Amoeba, plasma membrane; earthworm, body surface/'skin'; insect, tracheole; fish, lamellae; mammal, alveoli.

**13** This would result in excessive water loss and the plant would dehydrate. At night photosynthesis cannot occur and therefore there is no advantage to the plant of the stomatal pores remaining permanently open.

**14** The movement of water and mineral ions through the xylem is a passive process and therefore the cells do not need to be alive. The movement of organic solutes involves active transport and therefore requires ATP. The production of ATP, via respiration, can only occur in living cells.

**15** Water moves through the cell walls in the apoplast pathway and through the cytoplasm and plasmodesmata in the symplast pathway. The apoplast pathway is blocked at the endodermis.

**16** Any two of the following: A decrease in environmental temperature; an increase in humidity; a decrease in wind speed; a decrease in light intensity.

**17** False; they have a thinner cuticle than mesophytes as water loss is not a problem, however the waxy cuticle also helps to prevent the surface of the leaves from becoming waterlogged.

**18** If the xylem is damaged, the plant will die due to lack of water.

**19** Water is transported in the xylem in one direction, up the stem from the roots to the leaves; Sucrose is transported in the phloem both up and down the stem from 'source' to 'sink'.

**20 a** radius = 0.04 cm, therefore volume = $3.14 \times 0.04^2 \times 77 = 0.387 \, cm^3 s^{-1}$
  **b** $0.387 \times 60 \times 60 = 139 \, cm^3 h^{-1}$

**21** Gas exchange occurs directly with the tissue cells, via the tracheoles.

**22** They supply the cardiac muscle with blood that is rich in oxygen and glucose.

**23** The thicker muscular wall in the left ventricle generates higher pressure so that blood can travel further distances along the systemic circulation. Blood leaving the right ventricle only travels a short distance to the lungs and therefore does not need to be at such a high pressure.

**24** During ventricular systole all the blood leaves the ventricle, therefore during ventricular diastole the pressure falls to 0 kPa; however during ventricular diastole the semi-lunar valve closes, preventing back-flow of blood into the ventricle and maintaining a relatively high pressure in the aorta.

**25** The delay allows time for the ventricles to fill with blood before ventricular systole.

**26** Time taken for one cardiac cycle (i.e. from R to R) = 0.4 s. Heart rate = 60/0.4 = 150 bpm.

**27** During periods of high metabolic activity, e.g. exercise or an animal escaping a predator, blood needs to be redistributed around the body. Vasodilation occurs to increase blood flow to the skeletal muscles and to the skin surface (to lose excess heat); vasoconstriction occurs to reduce blood flow to unnecessary organs, e.g. the gut.

**28** Blood will flow from left to right; valve X will be closed and valve Y will be open.

**29** At the venule end of the capillary the hydrostatic pressure is lower than the osmotic pressure, so water moves back into the capillary by osmosis. Some of the water is not returned to the capillary, and *excess* tissue fluid drains into blind-ended lymph capillaries that return it to the blood via the thoracic duct.

**30** Myoglobin has a very high affinity for $O_2$ and acts as a store of $O_2$. It will only release its $O_2$ at a very low $pO_2$, but it will release it all, which allows the muscle fibres to continue to respire aerobically (aerobic respiration is far more efficient than anaerobic respiration).

**31** The $H^+$ ions need to be buffered by haemoglobin, and therefore they cause the oxyhaemoglobin to dissociate. This is important as it allows the released $O_2$ to diffuse into the body tissues for respiration.

**32** Intracellular digestion involves enzymes hydrolysing molecules within cells. Extracellular digestion involves the secretion of enzymes from cells into a cavity/gut where hydrolysis reactions occur. (Saprophytic organisms, such as fungi and bacteria, also carry out extracellular digestion by secreting enzymes into the environment to digest dead organic matter.)

**33** Mechanical digestion increases the overall surface area of the food that is consumed; this increases the rate of chemical digestion by enzymes, making digestion more efficient.

**34** Endopeptidases produce more 'free ends' for the action of exopeptidases, this increases the efficiency of protein digestion. If only one type of enzyme was released protein digestion would be slower and may be incomplete.

**35** Drains excess tissue fluid and returns it to the blood.

**36** Table showing dentition of a fox and a sheep

| Teeth | Fox | Sheep |
| --- | --- | --- |
| Incisors | Sharp | Small and flat topped — found in lower jaw only |
| Canines | Large and backward facing | Indistinguishable from the incisors |
| Diastema | Absent | Present |
| Premolars and molars | Modified with sharp cutting edges, and called carnassials | Interlocking with a large surface area, they also have sharp enamel ridges |

**37** So that the bacteria are not killed by the acidic pH in the stomach and they are provided with the optimum pH for their enzymes.

**38** By ensuring that pork is cooked thoroughly and by preventing the release of untreated sewage onto farmland.

# Index